Wine Growing Essentials for Small Scale Sustainable Viniculture

Richard Skiba

AFTER MIDNIGHT
PUBLISHING

Skiba, Richard (author)

Wine Growing Essentials for Small Scale Sustainable Viniculture

ISBN 978-1-7638440-1-8 (Paperback) 978-1-7638440-2-5 (eBook) 978-1-7638440-3-2 (Hardcover)

Non-fiction

Contents

Chapter 1
Introduction to Small-Scale Sustainable Viniculture

The Essence of Viticulture and Viniculture

Viticulture, derived from the Latin term *vitis cultura* meaning "vine-growing," is the science and art of cultivating and harvesting grapes. It is also referred to as viniculture (*vinis cultura*, "wine-growing") or winegrowing, emphasizing its role in the production of grapes for winemaking. As a branch of horticulture, viticulture focuses on the development and care of grapevines, particularly the species *Vitis vinifera*, the common grapevine. Native to regions stretching from Western Europe to the Persian shores of the Caspian Sea, *Vitis vinifera* has proven to be an exceptionally adaptable plant. This adaptability has allowed grapevines to thrive in diverse climates and geographies, leading to the establishment of viticulture on every continent except Antarctica.

A viticulturist is responsible for the comprehensive management of vineyards, ensuring that grapevines produce high-quality fruit suitable for winemaking or other purposes. Key duties of a viticulturist include monitoring and controlling pests and diseases that can harm the vines, applying fertilizers to optimize growth, and managing irrigation to maintain proper water balance. Canopy management is another critical task, involving the careful arrangement of vine leaves and shoots to optimize sunlight exposure, air circulation, and grape development.

Additionally, viticulturists monitor the fruit as it matures, paying close attention to its sugar content, acidity, and other characteristics that influence the quality of the final wine. Harvest timing is a major decision in the viticulture process, as it significantly affects the flavour, aroma, and structure of the resulting wine. After harvest, viticulturists also focus on vine pruning during the winter months to prepare the plants for the next growing cycle, ensuring balanced growth and yield.

Viticulturists often work closely with winemakers, as the quality of the grapes is foundational to the winemaking process. Effective vineyard management directly impacts the flavour profile, aroma, and overall characteristics of the wine. This collaboration ensures that vineyard practices align with the desired outcomes in winemaking, from crafting bold reds to delicate whites.

Over time, a wide variety of grapes has been cultivated for viticulture, with numerous varieties approved in the European Union as true winegrowing grapes. These varieties represent a rich genetic diversity, allowing for the production of wines with distinct regional identities. From classic varieties like Cabernet Sauvignon and Chardonnay to lesser-known indigenous grapes, viticulture embraces this diversity to cater to evolving consumer preferences and environmental conditions.

Viticulture is a dynamic and multifaceted field that combines scientific precision with agricultural artistry. Its global reach, coupled with the intricate role of the viticulturist in managing every stage of grape development, underscores its importance in the agricultural and winemaking industries. As viticulture continues to adapt to new environments and challenges, its role in shaping the world of wine remains as vital as ever.

Viticulture and viniculture are two interrelated yet distinct fields within the broader context of grape cultivation and winemaking. Understanding their differences is essential for appreciating their respective roles in the production of grapes and wine.

Viticulture, derived from the Latin term *vitis cultura*, refers to the cultivation and management of grapevines. This discipline encompasses a wide range of agricultural practices aimed at growing grapes for various purposes, including winemaking, fresh consumption, and the production of raisins and juice. The focus of viticulture is on the science and techniques necessary to cultivate healthy grapevines, optimize grape yield and quality, and maintain vineyard ecosystems [1, 2]. In contrast, viniculture, from the Latin *vinis cultura*, specifically pertains to the cultivation of grapes intended for winemaking. As a subset of viticulture, viniculture concentrates on practices that directly influence wine production, such as grape variety selection and harvest timing, which are critical for enhancing the wine's quality and flavour [3].

The activities involved in viticulture are extensive and include soil preparation, vineyard design, pest and disease control, and canopy management, among others. These practices apply to any grape-growing endeavour, making viticulture a comprehensive field [1, 4]. Conversely, viniculture is more specialized, focusing solely on grape varieties used for winemaking. It involves adapting vineyard practices to enhance specific wine characteristics, such as tannins, sugar levels, and acidity. Key activities in viniculture include determining optimal harvest times and collaborating with winemakers to align vineyard management with wine production goals [3, 5].

Richard Skiba

Figure 1: Cabernet Sauvignon, prior to harvest, at Laurel Bank Vineyard, Tasmania, in Australia's most southerly wine-producing region. Laurel Bank Wines, CC BY 2.0, via Flickr.

Viticulture is recognized as a broader scientific discipline within horticulture, integrating agricultural science and vineyard ecology to study grapevine biology and cultivation techniques [1, 2]. Research in viticulture aims to improve grapevine health and sustainability, particularly in the face of climate change and environmental challenges [1, 5]. Viniculture, while overlapping with enology—the study of wine and winemaking—focuses on grape-growing decisions tailored to achieve specific winemaking outcomes. This specialized focus makes viniculture integral to producing high-quality wines [3, 5].

The goals of viticulture and viniculture reflect their distinct scopes. Viticulture aims to grow healthy, high-quality grapes efficiently and sustainably, adapting to various environments and purposes beyond wine production [1, 6]. In contrast, viniculture's goal is to produce grapes optimized specifically for winemaking, influencing factors such as taste, aroma, and balance to meet the standards required for quality wine production [3, 5].

An illustrative example can be drawn from a vineyard in California that cultivates both table grapes for fresh consumption and wine grapes for vinification. The overall management of both grape types falls under viticulture, while the specific practices used to optimize the Cabernet Sauvignon for premium wine production—such as adjusting harvest timing and managing canopy growth—are categorized as viniculture [4].

While viticulture encompasses the broader science of grape cultivation, viniculture represents a specialized branch focused on producing grapes specifically for winemaking. The interconnection between the two fields is evident, as viniculture relies on the foundational practices established in viticulture. However, their distinction lies in their focus and application: viticulture addresses the cultivation of grapes in general, whereas viniculture hones in on optimizing grapes for wine production. Together, they form the backbone of the grape-growing and winemaking industries, contributing to the rich tapestry of viticulture and viniculture practices worldwide.

A Brief History

The history of viticulture, spanning thousands of years, reflects the intricate relationship between humans and grape cultivation, evolving through various cultural, economic, and environmental contexts.

The origins of viticulture can be traced back to the Near East, particularly in the South Caucasus region, where archaeological findings indicate the earliest evidence of grape cultivation and winemaking. Traces of grape residue found in pottery suggest that wild grapes (Vitis vinifera sylvestris) were domesticated into cultivated varieties (Vitis vinifera vinifera) around this time [7]. The development of fermentation likely occurred accidentally when crushed grapes were exposed to natural yeasts, leading to the production of fermented beverages [7].

In ancient Mesopotamia and Egypt, grapes were cultivated for various purposes, including ceremonial and medicinal uses. The Sumerians and Egyptians not only grew grapes but also produced wine, as evidenced by wall paintings in Egyptian tombs that depict vineyards and winemaking activities [7]. The Phoenicians played an importantrole in spreading viticulture across the Mediterranean, introducing grape cultivation to regions such as Greece, Italy, and North Africa, thereby enhancing the cultural significance of wine [7].

Viticulture flourished in ancient Greece, where grapes were central to agricultural practices and wine was integral to social and religious life. The Greeks introduced systematic viticulture techniques, including pruning and trellising, which facilitated the expansion of wine trade throughout the Mediterranean [7]. The Roman Empire further revolutionized viticulture by establishing vineyards across its territories, including modern-day France, Spain, and Germany. They implemented advanced techniques such as soil analysis and grafting, which became essential after the phylloxera crisis in the 19th century [8]. Roman writers like Pliny the Elder documented viticulture practices, preserving valuable knowledge for future generations [7].

During the Middle Ages, monastic orders played a pivotal role in preserving and refining viticulture techniques, particularly in Europe. Monasteries became centres of viticultural innovation, especially in regions like Burgundy and Champagne in France, and the Rhine Valley in Germany [7]. The expansion of viticulture during this period was facilitated by trade and conquest, leading to the introduction of grape cultivation in Eastern Europe and the British Isles [7].

Prehistoric Beginnings

The earliest evidence of grape cultivation and winemaking comes from the South Caucasus region (modern-day Georgia), where archaeologists have found traces of grape residue in pottery. Wild grapes (Vitis vinifera sylvestris) were first domesticated into cultivated varieties (Vitis vinifera). Fermented beverages made from grapes likely arose accidentally when wild grapes were crushed and exposed to natural yeasts.

Ancient Civilizations

The Sumerians and Egyptians cultivated grapes and made wine for ceremonial, medicinal, and recreational purposes. Wall paintings in Egyptian tombs depict vineyards and winemaking, indicating its cultural importance. Phoenicians were instrumental in spreading viticulture around the Mediterranean, introducing grape cultivation to regions like Greece, Italy, and North Africa.

Greek and Roman Contributions

In Ancient Greece, grapes were central to agriculture, and wine played a vital role in social and religious life. Greeks introduced systematic viticulture techniques and expanded wine trade throughout the Mediterranean. The Romans revolutionized viticulture, establishing vineyards across their empire and introducing advanced techniques such as soil analysis, grafting, and vineyard zoning. Roman writers like Pliny the Elder documented viticulture practices, preserving knowledge for future generations.

The Middle Ages

Monasteries preserved and refined viticulture techniques, particularly in Europe, during the Dark Ages. Vineyards flourished under religious orders, particularly in France (e.g. Burgundy, Champagne) and Germany (e.g. the Rhine Valley). Viticulture spread to new regions through trade and conquest, including parts of Eastern Europe and the British Isles.

The Age of Exploration and Colonization

European colonists brought grapevines to the Americas, South Africa, and Australia. Spanish missionaries established vineyards in the Americas, including California and South America. Improved understanding of grape diseases and soil management emerged during this period. French viticulture reached new heights, with regions like Bordeaux and Burgundy becoming renowned for their wines.

The Industrial Revolution and Modernization

A devastating outbreak of the phylloxera pest nearly destroyed European vineyards in the mid-19th century, leading to the development of resistant rootstocks through grafting with American vine species. Mechanization of viticulture, including plows and presses, began to replace manual labor. Fermentation processes improved with advancements in microbiology, particularly through the work of Louis Pasteur.

The 20th Century: Globalization and Innovation

The Prohibition era in the United States (1920–1933) severely impacted American viticulture but also led to innovations in table grape production. Regions like California, Australia, Chile, and South Africa emerged as major players in global viticulture, challenging European dominance. Precision agriculture, irrigation technologies, and advancements in plant breeding revolutionized vineyard management. The introduction of controlled environments and disease-resistant varieties improved grape quality and yields.

The 21st Century: Sustainability and Global Challenges

Shifting climates have forced viticulturists to adapt with new varieties and vineyard practices. Organic, biodynamic, and regenerative viticulture have gained prominence to address environmental concerns. The globalization of wine has brought attention to lesser-known regions, creating new opportunities for small-scale and boutique growers.

-6000 -3000 -1000 500 1500 1800 1900 2000

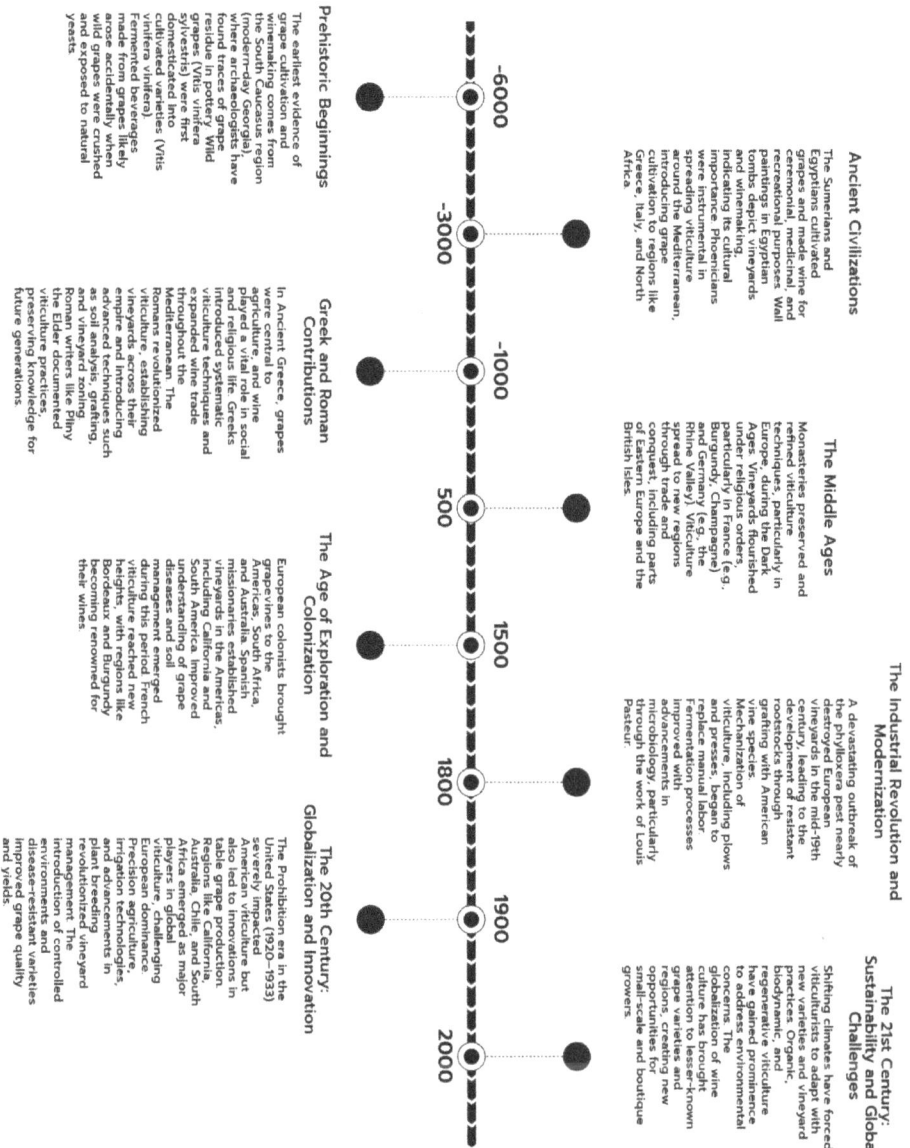

Figure 2: Timeline of history of viticulture.

7

The Age of Exploration saw European colonists introducing grapevines to the Americas, South Africa, and Australia. Spanish missionaries established vineyards in California and South America, significantly impacting the global viticulture landscape [7]. This period also marked advancements in scientific understanding related to grape diseases and soil management, with French viticulture reaching new heights in renowned regions like Bordeaux and Burgundy [7].

The mid-19th century was marked by the devastating phylloxera crisis, which nearly destroyed European vineyards. This crisis prompted the development of resistant rootstocks through grafting with American vine species, a practice that remains integral to modern viticulture [8]. Technological advancements during this period included the mechanization of viticulture and improvements in fermentation processes, largely influenced by the work of Louis Pasteur in microbiology [7].

The Prohibition era in the United States (1920–1933) had a profound impact on American viticulture, leading to innovations in table grape production (Visconti, 2024). The emergence of New World wines from regions like California, Australia, Chile, and South Africa challenged the traditional dominance of European wines, fostering a global wine culture that continues to evolve [7]. Technological innovations, including precision agriculture and advancements in plant breeding, have revolutionized vineyard management, enhancing grape quality and yields [7].

Today, viticulture faces significant challenges related to climate change, necessitating adaptations in vineyard practices and grape varieties [7]. Sustainable practices, including organic and biodynamic viticulture, have gained prominence as growers seek to address environmental concerns and enhance soil health [7]. The globalization of wine culture has also led to increased demand for diverse grape varieties and sustainable production methods, creating opportunities for small-scale and boutique growers [7].

Wines and Grapes

Wine Producing Environment

The grapevine (Vitis vinifera) is a highly specialized plant that thrives in specific environmental conditions, primarily found in temperate latitudes between 30° and 50° in both hemispheres. These regions typically experience annual mean temperatures ranging from 10°C to 20°C (50°F to 68°F), which are conducive to grape cultivation. The climate in these areas is significantly influenced by geographical features such as large bodies of water and mountain ranges. These features help to moderate temperature fluctuations by storing heat during the day and releasing it at night, thus protecting the vines from extreme temperature drops [9, 10]. This natural regulation is important for the consistent development of grapes, particularly in regions where temperature variability is common [11].

The structure of the grapevine is composed of several essential parts that contribute to its growth and productivity. The root system, including fender roots, anchors the plant and absorbs water and nutrients while also storing food [12]. The trunk serves as the central support, branching into arms that facilitate the transport of nutrients and water to the upper parts of the plant. Higher up, the vine features nodes, internodes, leaves, clusters of grapes, and tendrils, which play a vital role in supporting the vine by attaching it to surrounding structures [13]. This intricate structure allows the grapevine to adapt efficiently to its environment, enhancing its productivity and resilience in viticulture [14].

Grapes themselves are classified as berries and grow in clusters, which can vary in compactness. The arrangement of these clusters affects harvesting efficiency, as some grape species ripen collectively while others do so individually [15]. Each grape berry is attached to the rachis via a pedicel, which serves as a conduit for water and nutrients. The skin of the grape, accounting for 5–20% of its total weight, contains tannins and aromatic substances that are essential for winemaking, influencing the colour and flavour during processes such as colour extraction and aroma dissolution [16, 17]. Tannins are particularly important during the ripening phase, contributing to the formation of colour and body structure in the grapes [18].

The concept of terroir, which encompasses climate, slope, and soil, is critical for grapevine growth and the quality of grapes produced in a vineyard. Climate is the most influential factor, as each grape variety thrives within specific climatic ranges [9, 11]. Grapes require approximately 1,300–1,500 hours of sunshine during the growing season and about 690 mm (27 inches) of annual rainfall, ideally concentrated in winter and spring to avoid harvest-related issues such as fungal diseases [10, 19]. Seasonal conditions, particularly summer temperatures averaging 22°C (72°F) and winter temperatures around 3°C (37°F), are vital for optimal ripening and vine health [18].

Slope also plays a significant role in viticulture, as sloped terrains enhance sunlight exposure and improve drainage, reducing the risk of overly moist soil that can harm the roots [14, 16]. In cooler northern hemisphere regions, south-facing slopes are preferred for their increased sunlight, while north-facing slopes are advantageous in warmer areas [20, 21]. Soil quality is equally important, with well-aerated, loose-textured soils that provide good drainage being optimal for healthy root systems [7]. Poorly drained soils can restrict root growth, negatively impacting vine vigour and fruit yield [11, 12].

The grapevine is a highly specialized plant adapted to thrive in specific environmental conditions. Its structure supports efficient nutrient absorption and fruit production, while the characteristics of its berries are essential to winemaking. The interplay of climate, slope, and soil defines the unique qualities of each vineyard, influencing the flavour, aroma, and quality of the resulting grapes. Understanding these factors allows viticulturists to optimize vineyard practices and produce grapes that reflect the distinct terroir of their growing region [22].

Wine Naming Conventions

The naming of wine types is influenced by a variety of factors, including geographical location, grape varieties, production methods, and historical traditions. These naming conventions provide consumers with key information about the wine's origin, style, and characteristics. Below are the primary ways in which wine types are named:

By Geographic Origin (Old World Tradition): In many European countries, wines are named after the region or appellation where the grapes are grown and the wine is produced. This practice emphasizes the concept of *terroir*, or the unique combination of climate, soil, and geography that influences the wine's character. Often, the grape variety is not explicitly stated, as the region itself implies the grapes used.

- **Examples**:
  - **Bordeaux** (France): Refers to wines produced in the Bordeaux region, often blends of Cabernet Sauvignon, Merlot, and other varieties.
  - **Chianti** (Italy): Named after the Chianti region in Tuscany, typically made from Sangiovese grapes.
  - **Rioja** (Spain): Named after the Rioja region, predominantly made with Tempranillo grapes.

- o **Champagne** (France): Sparkling wine exclusively from the Champagne region, primarily made with Chardonnay, Pinot Noir, and Pinot Meunier grapes.

This system is most common in "Old World" wine regions such as France, Italy, Spain, and Germany, where strict regulations govern the types of grapes and production methods allowed in each region.

By Grape Variety (New World Tradition): In the "New World" wine regions, such as the United States, Australia, and South Africa, wines are often named after the grape variety used to produce them. This approach highlights the grape's role in determining the wine's flavour and style, making it easier for consumers to identify and understand the wine.

- **Examples**:

    - o **Cabernet Sauvignon**: A red wine named after the grape variety, known for its bold flavours and aging potential.

    - o **Chardonnay**: A white wine named after the grape, ranging from crisp and unoaked to rich and buttery styles.

    - o **Sauvignon Blanc**: A white wine known for its high acidity and herbaceous notes.

For a wine to be labelled by grape variety in many countries, regulations typically require that a certain percentage (often 75–85%) of the wine be made from that grape.

By Production Method or Style: Some wines are named based on the method of production or their style, which can include the winemaking process, the type of fermentation, or the residual sugar content.

- **Examples**:

    - o **Sparkling Wine**: Refers to wines with carbonation, such as Champagne (France), Prosecco (Italy), and Cava (Spain).

    - o **Rosé**: A wine style with a pink hue, made by limiting the contact between grape skins and juice during fermentation.

    - o **Dessert Wine**: Wines with high residual sugar content, such as Port, Sauternes, or Ice Wine.

- o **Fortified Wine**: Wines with added distilled spirits, such as Sherry, Madeira, or Port.

By Blends or Proprietary Names: Blended wines or those with unique branding are sometimes given proprietary names by the producer, rather than naming them after a grape variety or region.

- **Examples**:

  - o **Super Tuscans**: High-quality blends from Tuscany that do not conform to traditional Italian wine regulations, often named creatively (e.g., Sassicaia, Tignanello).

  - o **Opus One**: A premium Napa Valley wine with a proprietary name, made from a Bordeaux-style blend.

Producers may use proprietary names to market unique blends or emphasize the wine's exclusivity.

By Legal Designations and Classifications: Many countries use strict legal frameworks to regulate wine naming. These systems often include designations of origin, quality levels, and classifications.

- **Examples**:

  - o **AOC/AOP (France)**: "Appellation d'Origine Contrôlée/Protégée" indicates wines that meet strict regional guidelines (e.g., Chablis, Margaux).

  - o **DOC/DOCG (Italy)**: "Denominazione di Origine Controllata/Controllata e Garantita" indicates high-quality wines from specific regions (e.g., Barolo, Brunello di Montalcino).

  - o **AVA (USA)**: "American Viticultural Area" names wines by their region of origin (e.g., Napa Valley, Willamette Valley).

By Historical or Cultural Traditions: Some wines retain names rooted in history or local culture, reflecting traditions that have been passed down for generations.

- **Examples**:

  - o **Malbec** (Argentina): A grape variety traditionally associated with Cahors in France but has become synonymous with Argentine wine.

- o **Amarone** (Italy): A specific style of rich, dry red wine made from dried grapes in the Veneto region.

Wine naming conventions provide vital information about the wine's origins, grape composition, and production methods. Old World wines focus on regional identities and traditional practices, while New World wines prioritize grape variety and stylistic transparency. Understanding these naming systems helps consumers navigate the vast world of wine with confidence and appreciation for its diversity.

Types of Wine and Grapes: Understanding Their Relationship

The type of wine produced is deeply influenced by the grape varieties used, as grapes provide the fundamental flavours, aromas, and characteristics of the wine. Each type of wine has unique profiles based on the grape variety, growing conditions, and winemaking techniques. Following is an exploration of the major types of wine, the grape varieties that contribute to them, and how they relate.

1. Red Wines

Red wines are made from dark-skinned grape varieties. The skins, seeds, and stems are often left in contact with the juice during fermentation, imparting colour, tannins, and structure to the wine.

Red wines are primarily produced from dark-skinned grape varieties, which play a significant role in defining the wine's characteristics. The process of maceration, where grape skins, seeds, and sometimes stems are left in contact with the juice during fermentation, is fundamental to this production. This interaction not only imparts colour to the wine through anthocyanins but also contributes tannins and other phenolic compounds that enhance the wine's structure and complexity. The duration of skin contact directly influences the wine's colour intensity and tannin levels, which are critical for the wine's aging potential and overall mouthfeel [23-25].

Key Grape Varieties:

- **Cabernet Sauvignon**: Known as the "king of red grapes," it produces full-bodied wines with bold flavours of blackcurrant, cedar, and spices. Often used in blends (e.g., Bordeaux).

- **Merlot**: A softer, fruit-forward grape with flavours of plum, cherry, and chocolate. Frequently blended with Cabernet Sauvignon.

- **Pinot Noir**: Produces light- to medium-bodied wines with flavours of red berries, earth, and mushrooms. Famous in Burgundy and Oregon.

- **Syrah/Shiraz**: Known for bold, spicy wines with flavours of blackberry, pepper, and smoke. Syrah is used in Rhône Valley blends, while Shiraz is popular in Australia.

- **Zinfandel**: Grows primarily in California, offering jammy, spicy wines with blackberry and pepper notes.

Key grape varieties such as Cabernet Sauvignon, Merlot, Pinot Noir, Syrah/Shiraz, and Zinfandel each contribute distinct flavours and characteristics to red wines. Cabernet Sauvignon, often termed the "king of red grapes," is celebrated for its full-bodied profile and robust flavours, making it a staple in Bordeaux blends [26, 27]. Merlot, known for its softer and fruit-forward nature, often complements Cabernet Sauvignon in blends, adding roundness and approachability [28]. In contrast, Pinot Noir is recognized for its delicate and nuanced wines, which are lighter in body and often exhibit earthy and fruity notes, particularly in cooler climates like Burgundy [29, 30]. Syrah, or Shiraz, is noted for its bold and spicy flavour profile, thriving in warmer regions and contributing to rich, fruit-driven wines [27, 31]. Zinfandel, primarily cultivated in California, showcases a range of styles from light and fruity to robust and high in alcohol, reflecting its versatility [30, 32].

The grape skins play a vital role in red wine, as they provide the tannins and colour. The ripeness and growing conditions of the grapes (e.g., cooler climates for lighter wines, warmer climates for richer wines) also influence the wine's acidity, sweetness, and body.

Figure 3: The "cap" of grape skins that forms on a fermenting red wine being pushed down. Wollombi, CC BY 2.0, via Wikimedia Commons.

The grape skins are integral to the production of red wine, as they contain anthocyanins responsible for the wine's colour and tannins that provide structure and astringency. The phenolic compounds extracted during maceration not only affect the wine's sensory properties but also its aging potential [33, 34]. The ripeness of the grape skins at harvest significantly influences the wine's flavour profile; grapes from cooler climates tend to have higher acidity and lighter body, while those from warmer regions yield richer, fruitier wines with lower acidity [35, 36]. This interplay of climate and grape variety underscores the importance of terroir in winemaking, as it shapes the characteristics of the final product [37, 38].

Growing conditions, or terroir, profoundly influence the characteristics of red wines. Factors such as climate, soil composition, and vineyard management practices interact to determine the quality and style of the grapes. For instance, cooler climates favour varieties like Pinot Noir, resulting in wines with refined acidity and subtle complexity, while warmer regions excel in producing fuller-bodied wines from varieties such as Cabernet Sauvignon and Shiraz [28, 39, 40]. The balance of acidity, sweetness, and body in red wine is a direct reflection of these growing conditions, and winemakers must carefully assess grape ripeness and health during harvest to achieve the desired wine profile [41, 42].

2. White Wines

White wines are usually made from green-skinned or light-coloured grape varieties, with minimal skin contact during fermentation. They are typically lighter and more acidic than red wines. This method is fundamental as it results in the lighter body and pale colour characteristic of white wines, while also enhancing their crispness, acidity, and aromatic profile [43, 44]. The production techniques emphasize the purity of flavours, making white wines a refreshing choice for various occasions and food pairings, particularly when served chilled [45].

Key Grape Varieties:

- **Chardonnay**: A versatile grape that produces a wide range of styles, from crisp, citrusy wines to rich, buttery ones aged in oak. Famous in Burgundy and California.

- **Sauvignon Blanc**: Known for its high acidity and herbal, grassy notes, often combined with flavours of lime, green apple, and tropical fruits. Popular in New Zealand and the Loire Valley.

- **Riesling**: A highly aromatic grape producing wines ranging from bone-dry to sweet. Known for its floral, citrus, and petrol-like aromas, particularly in Germany and Alsace.

- **Pinot Grigio/Pinot Gris**: Produces light, crisp wines with notes of pear, apple, and lemon in Italy (Pinot Grigio) or richer, spicier wines in Alsace (Pinot Gris).

- **Chenin Blanc**: A versatile grape offering everything from sparkling to sweet wines, with flavours of honey, apple, and apricot.

The acidity and aromatic compounds in the grape largely define the freshness and flavour profile of white wines. For example, high-acid grapes like Sauvignon Blanc thrive in cooler climates, producing vibrant wines.

The diversity of white wines can be attributed to the various grape varieties used in their production, each contributing unique characteristics to the final product. For instance, Chardonnay is one of the most versatile white grape varieties, producing a wide range of styles depending on the climate and winemaking techniques. In cooler regions, such as Chablis in Burgundy, Chardonnay yields crisp, citrus-driven wines, while warmer climates like California produce richer, oak-aged versions with flavours of butter and tropical fruits [46, 47]. Similarly, Sauvignon Blanc, known for its high acidity and herbal notes, thrives in cooler climates, resulting in wines with vibrant flavours of lime and green apple, particularly noted in regions like the Loire Valley and New Zealand [48].

Figure 4: Chardonnay Grapes, SLO County Edna Valley. Harold Litwiler, CC BY-SA 2.0, via Flickr.

Riesling, another prominent white grape, is highly aromatic and can produce wines that range from bone-dry to sweet. Its flavour profile is characterized by floral and citrus notes, with a unique petrol-like aroma that develops with age, especially in wines from Germany and Alsace [49, 50]. Pinot Grigio, known as Pinot Gris in Alsace, offers two distinct styles: the light and crisp Italian version and the fuller-bodied, spicier Alsatian variant [51, 52]. Lastly, Chenin Blanc showcases remarkable versatility, producing everything from sparkling to dessert wines, with flavour notes that include honey and apricot, particularly celebrated in the Loire Valley and South Africa [53, 54].

The relationship between grape varieties and the resulting wines is significantly influenced by growing conditions. High-acid grapes like Sauvignon Blanc and Riesling flourish in cooler climates, where slow ripening preserves their natural acidity and enhances their aromatic qualities [55, 56]. In contrast, warmer climates tend to produce grapes with higher sugar levels and lower acidity, leading to fuller-bodied wines, as seen with Chardonnay [57]. The minimal skin contact during fermentation is essential in maintaining the light body and clean flavours of white wines, allowing the grape's natural aromas and acidity to shine through [58].

3. Rosé Wines

Rosé wines are made from red or black grape varieties, but with minimal skin contact during fermentation, resulting in a pink colour. They can range from dry to sweet. Rosé wines represent a unique and versatile category within the wine spectrum, bridging the gap between red and white wines. They are primarily produced from red or black grape varieties, utilizing minimal skin contact during fermentation to achieve their characteristic pink hue. This brief interaction with the grape skins allows for a delicate extraction of colour and flavour, resulting in a wine that is both approachable and refreshing, with lighter tannins compared to red wines [59]. The colour of rosé can vary significantly, from pale salmon to deep ruby, influenced by the grape variety and specific winemaking techniques employed [60]. Furthermore, rosé wines can be crafted to suit a wide range of palates, from bone-dry to sweet, making them suitable for various occasions and food pairings [61].

Key Grape Varieties:

- **Grenache**: Often used for dry rosés with notes of strawberry, watermelon, and herbs.

- **Syrah**: Contributes darker, spicier rosés.

- **Mourvèdre**: Adds structure and savory notes to rosé blends.

- **Pinot Noir**: Creates elegant, delicate rosés with flavours of red berries.

The grape variety determines the style of rosé. For example, Grenache-based rosés from Provence are known for their pale colour and light body, while Syrah-based rosés are darker and fuller.

Grenache is one of the most widely used grapes, particularly in Provence, where it contributes flavours of strawberry and watermelon, along with a light, crisp body [62]. Syrah, known for its boldness in red wine production, imparts deeper hues and richer flavours, such as blackberry and plum, to rosé wines [62]. Mourvèdre adds complexity with its earthy and herbal notes, enhancing the wine's structure and food-pairing versatility [62]. Pinot Noir, on the other hand, yields rosés that are elegant and aromatic, often featuring red berry notes complemented by floral undertones [62]. The choice of grape variety not only affects the flavour but also the acidity and body of the wine, influencing its overall character and suitability for different culinary experiences [61].

Figure 5: Syrah grapes growing in the Central Coast of California near Soledad. Hahn Family Wines, CC BY 2.0, via Wikimedia Commons.

Winemaking techniques significantly impact the final style of rosé wines. The most common method involves limited maceration, where grape skins are in contact with the juice for a short duration, typically a few hours to a day. This technique allows winemakers to control the extraction of colour and flavour [59]. Another method, known as saignée, involves "bleeding" off some juice from red wine production, which can result in a more concentrated and robust rosé [62]. Additionally, blending red and white wines is a less common practice, primarily seen in sparkling rosé production, such as in Champagne [62]. Each of these techniques contributes to the diversity of rosé wines, allowing for a wide range of styles and flavour profiles that cater to different consumer preferences [61].

4. Sparkling Wines

Sparkling wines are characterized by their effervescence, which is primarily due to the carbon dioxide produced during fermentation. This effervescence not only defines the sensory experience of sparkling wines but also contributes to their celebratory nature. The production of sparkling wines encompasses a variety of styles, ranging from bone-dry to sweet, influenced by grape varieties, production methods, and regional characteristics. The versatility of sparkling wines makes them suitable for both casual and formal occasions, with many prestigious labels steeped in tradition and history [63, 64]. They can be made from a variety of grapes using traditional or tank fermentation methods.

Key Grape Varieties:

- **Chardonnay**: A key grape in Champagne, known for its elegance and acidity.

- **Pinot Noir**: Provides body and structure to sparkling wines, often used in Champagne blends.

- **Pinot Meunier**: Adds fruitiness to Champagne blends.

- **Prosecco Grapes (Glera)**: The primary grape in Prosecco, offering light, fruity, and floral flavours.

The grape's acidity and sugar levels are essential for sparkling wine production, as they must balance the wine's effervescence and freshness. Cooler climates, like those in Champagne, are ideal for high-acid grape varieties.

The choice of grape varieties is crucial in determining the flavour profile and structure of sparkling wines. Chardonnay, for instance, is renowned for its elegance and high acidity, making it a staple in many sparkling wines, particularly from the Champagne region. It contributes crispness and finesse, often showcasing citrus and green apple notes [63]. Pinot Noir, a red grape, adds body and complexity, with its subtle red fruit flavours enhancing the overall profile of blends. When used alone, it can produce vibrant rosé sparkling wines [63, 64]. Pinot Meunier, while often overshadowed, plays a significant role by adding fruitiness and approachability, making it essential for creating balanced blends [63, 64].

Figure 6: A cluster of the Pinot Noir grape variety. Ursula Brühl, Julius Kühn-Institut (JKI), Federal Research Centre for Cultivated Plants, Institute for Grapevine Breeding Geilweilerhof - 76833 Siebeldingen, GERMANY, CC BY-SA 4.0, via Wikimedia Commons.

Glera, the primary grape for Prosecco, offers a light and fruity profile, characterized by flavours of pear and green apple. This grape's easy-drinking nature has made Prosecco a popular choice for casual settings [63, 64]. The selection of these grape varieties is not arbitrary; they are chosen for their ability to maintain high acidity and balanced sugar levels, which are vital for the production of sparkling wines that are both refreshing and complex [63, 64].

The production of sparkling wines relies heavily on the acidity and sugar levels of the grapes used. High acidity is essential as it balances the sweetness and enhances the refreshing qualities of the wine. Cooler climates, such as those found in Champagne and northern Italy, are ideal for cultivating high-acid grape varieties, allowing for slow ripening that preserves acidity while developing nuanced flavours [63, 64]. Sugar levels are equally critical, particularly during the secondary fermentation process, where additional sugar and yeast are introduced to create the characteristic bubbles of sparkling wines. The initial sugar content must be carefully managed to achieve the desired style, from dry Brut to sweeter Demi-Sec and Doux wines [63, 64].

The production methods for sparkling wines significantly influence their flavour and texture. The Traditional Method (Méthode Traditionnelle) involves a second fermentation in the bottle, where the wine ages on its lees for an extended period. This method is known for producing complex flavours, including notes of brioche and toast, alongside fresh fruit characteristics [63, 64]. In contrast, the Tank Method (Charmat Method) is employed for wines like Prosecco, where the second fermentation occurs in large stainless-steel tanks. This method emphasizes freshness and fruit-forward flavours, resulting in lighter and more approachable wines [63, 64].

Research has shown that the production method can affect the sensory profile of sparkling wines. For instance, wines produced via the Traditional Method tend to exhibit a more complex character due to the extended aging on yeast, while those made using the Tank Method are often perceived as sweeter and fruitier [64]. The choice of method thus plays a pivotal role in defining the final product's quality and consumer acceptance [64].

Climate is a critical factor in the production of high-quality sparkling wines. Cooler climates help preserve the natural acidity of grapes, which is essential for creating crisp and refreshing wines. In regions like Champagne, the cool temperatures slow the ripening process, allowing grapes to develop complex aromas and flavours without sacrificing acidity [63, 64]. Conversely, warmer regions may lead to grapes ripening too quickly, resulting in lower acidity and higher sugar levels, which can compromise the balance necessary for sparkling wines [63]. However, advancements in vineyard management and winemaking techniques have enabled producers in warmer climates to create high-quality sparkling wines [63, 64].

5. Dessert Wines

Dessert wines represent a luxurious and diverse category of wines characterized by their sweetness and richness, making them ideal for pairing with desserts or enjoying independently after a meal. The sweetness of these wines is achieved through various specialized grape-growing and winemaking techniques, which result in concentrated flavours and complex aromas. The balance of sweetness, acidity, and flavour depth is important to the quality of dessert wines, allowing for a wide range of styles that cater to different palates and occasions [65, 66].

Key Grape Varieties:

- **Sémillon**: Used in Sauternes for botrytized dessert wines with honey and apricot notes.

- **Muscat**: Produces sweet wines with floral and grapey aromas.

- **Riesling**: Makes exceptional late-harvest and ice wines with flavours of honey, citrus, and peach.

- **Chenin Blanc**: Used in sweet wines from the Loire Valley.

The sugar content of the grapes at harvest is vital for dessert wines. Techniques like late harvesting or encouraging noble rot concentrate the sugars, creating rich, sweet wines.

The production of dessert wines involves several key techniques that enhance the natural sugar content of the grapes. One prominent method is late harvesting, where grapes are left on the vine longer than usual to over-ripen, concentrating their sugars and flavours. This technique often results in wines with intense sweetness and fruit-forward profiles, showcasing notes of tropical and stone fruits [65, 66].

Another significant method is the use of botrytized grapes, which involves the infection of grapes by the fungus *Botrytis cinerea*, commonly referred to as noble rot. This fungus dehydrates the grapes, concentrating their sugars and acids, leading to wines with a luscious texture and complex flavours, such as honey and apricot [67]. Notable examples include Sauternes from France, which exemplifies the richness that botrytized grapes can impart [67].

Figure 7: Botrytis cinerea on Riesling grapes, noble rot. Photographer: Tom MaackNo machine-readable author provided. T.o.m.~commonswiki assumed (based on copyright claims)., CC BY-SA 3.0, via Wikimedia Commons.

Ice wine production is another unique technique, where grapes are left on the vine until they freeze naturally. The frozen grapes are then harvested and pressed, yielding highly concentrated juice that is prized for its purity and vibrant flavours. Ice wines typically exhibit a harmonious balance of sweetness and acidity, making them particularly sought after in regions like Canada and Germany [65, 66].

Certain grape varieties are particularly well-suited for dessert wine production due to their ability to retain acidity while developing rich flavours as their sugars concentrate. Sémillon is a cornerstone of botrytized dessert wines, especially in Sauternes, where its natural affinity for noble rot results in wines with honeyed sweetness and complex notes of stone fruits [67].

Muscat is another important variety, known for its floral and grapey aromas, producing a range of sweet wines from light and refreshing to rich and syrupy. Examples include fortified Muscat wines from Rutherglen, Australia, and aromatic Moscato d'Asti from Italy [65, 66].

Riesling is celebrated for its versatility, excelling in late-harvest and ice wine styles. Its high natural acidity balances sweetness, resulting in wines with flavours of honey, citrus, and peach, particularly noted in German and Canadian dessert wines [65, 66]. Lastly, Chenin Blanc shines in sweet wines from the Loire Valley, contributing vibrant acidity and flavours of honey and dried fruits, making it a popular choice for long-lived dessert wines [65, 66].

The sugar content of grapes at harvest is critical in producing high-quality dessert wines. Techniques such as late harvesting, noble rot, and freezing naturally concentrate the sugars, leading to the characteristic richness of these wines. However, sweetness alone does not define a great dessert wine; acidity plays an equally vital role in providing balance and preventing the wine from feeling overly sweet or cloying [65, 66].

For instance, in Sauternes, the Sémillon grape's ability to develop noble rot enhances its sweetness while preserving vibrant acidity, creating a well-rounded wine. Similarly, Riesling-based ice wines benefit from the grape's high acidity, which offsets concentrated sugars, resulting in a sweet yet refreshingly crisp profile [65, 66]. Thus, the choice of grape variety, combined with specific methods of sugar concentration, ultimately determines the wine's final flavour profile, texture, and style.

6. Fortified Wines

Fortified wines have distilled spirits added, increasing their alcohol content and longevity. This process typically results in an alcohol content ranging from 15% to 22%, which not only preserves the wine but also contributes to its complexity and longevity [68, 69]. Historically, fortification was developed to ensure that wines could withstand long journeys, a necessity that has evolved into a hallmark of celebrated wine styles like Port, Sherry, and Madeira. These wines are renowned for their rich flavours, ranging from sweet and luscious to dry and nutty, showcasing a remarkable versatility that appeals to a broad spectrum of palates [68, 69].

Key Grape Varieties:

- **Touriga Nacional**: A key grape in Port, offering dark fruit and floral notes.

- **Palomino**: Used in Sherry, producing dry wines with nutty flavours.

- **Tinta Roriz (Tempranillo)**: Often used in fortified wines from Spain and Portugal.

The grapes used for fortified wines are selected for their ability to retain intense flavours and sugar levels during the fortification process.

The selection of grape varieties is significant in the production of fortified wines, as these grapes must possess the intensity and structure necessary to endure the fortification process. Touriga Nacional, for instance, is a flagship grape for Port production, known for its concentrated dark fruit flavours and floral aromas, which contribute significantly to the wine's depth and aging potential [70, 71]. Similarly, Palomino, the primary grape for Sherry, yields dry wines with delicate nutty and saline notes, particularly suited for the oxidative aging processes that define styles like Amontillado and Oloroso [68, 69]. Tinta Roriz, also known as Tempranillo, adds structure and earthy flavours, enhancing the complexity of both dry and sweet fortified wines [68, 72].

The fortification process itself is pivotal in determining the final character of the wine. For sweet fortified wines like Port, the spirit is added during fermentation, halting the process and leaving residual sugars that create a rich flavour profile [68, 70]. In contrast, dry styles like Sherry undergo fortification after fermentation, allowing for a different aging process that can impart unique characteristics through biological aging under flor yeast or oxidative aging in barrels [68, 69]. This careful manipulation of the winemaking process not only enhances the wine's longevity but also allows for the development of complex flavours over time [69, 70].

The relationship between grape selection and the resulting fortified wine is intricate. For sweet wines, high-sugar varieties like Touriga Nacional are essential, ensuring a balanced sweetness even after fortification [68, 69]. In dry fortified wines, the neutral profile of Palomino allows the aging process to shine, showcasing the complex nutty and savory notes that develop [68, 69]. The structural integrity and acidity of the grapes are also vital, as they help balance the sweetness and alcohol content, creating a harmonious drinking experience [68, 70].

Fortified wines encompass a diverse array of styles, each reflecting the traditions and techniques of their regions. Port, originating from Portugal, is celebrated for its rich sweetness and comes in various styles, including Ruby, Tawny, and Vintage [69, 70].

Sherry, from Spain, ranges from dry to sweet, with Palomino grapes dominating the production of dry styles, while sweeter varieties often include Moscatel or Pedro Ximénez [68, 69]. Madeira, produced on the Portuguese island, is known for its resilience and unique aging process, which enhances flavours of caramel, nuts, and dried fruit, resulting in a wide range of styles from dry to sweet [68-70].

Why Small-Scale and Sustainable?

The growing interest in small-scale and sustainable viticulture is driven by a combination of environmental, economic, social, and market-related factors. These motivations reflect both the challenges faced by the wine industry and the evolving preferences of consumers and producers. Environmental concerns play a significant role, as conventional viticulture often leads to environmental degradation through excessive water use, chemical inputs, and carbon emissions. Small-scale sustainable practices address these issues by emphasizing organic methods, reduced water usage, and carbon-efficient processes. Climate change has also forced vineyards to adapt to shifting growing conditions, making sustainability essential for long-term viability. Additionally, sustainable viticulture promotes practices like cover cropping, composting, and minimal soil disturbance, protecting soil health and encouraging biodiversity within the vineyard ecosystem. These methods contribute to healthier vines and more resilient vineyards.

Economic viability is another driver of small-scale and sustainable viticulture. Large-scale monoculture vineyards often require significant investments in machinery, chemicals, and labour. In contrast, small-scale sustainable vineyards can reduce costs by utilizing natural pest control, renewable energy, and minimal intervention methods. These producers often focus on quality over quantity, emphasizing premium, high-quality wines rather than mass production. Sustainable practices enhance wine quality by fostering healthier vines and better grape characteristics. Small-scale operations are also more agile, allowing producers to experiment with innovative methods like biodynamic farming or permaculture, which can yield unique and marketable wine profiles.

Social and ethical drivers further support the movement toward small-scale sustainable viticulture. Modern consumers increasingly demand transparency in how their products are made, valuing fair labour practices and environmentally friendly methods. Small-scale viticulture aligns with these ethical concerns, appealing to buyers seeking socially responsible options. Additionally, this approach often preserves local traditions, heritage grape varieties, and unique terroirs, contributing to cultural sustainability in wine-producing regions.

The market for small-scale and sustainable viticulture is expanding, fuelled by trends in consumer preferences, regulatory pressures, and the evolving wine industry. Consumers are showing a growing preference for premium and artisanal wines that offer unique flavour profiles and stories behind the bottle. Small-scale vineyards are ideally positioned to meet this demand. Eco-conscious buyers also seek wines labelled as organic, biodynamic, or sustainably produced, with certifications like USDA Organic or Demeter Biodynamic adding market appeal. Furthermore, the focus on natural and organic products extends to wine, with health-conscious consumers favouring wines made with minimal chemical inputs and fewer additives.

Niche and direct-to-consumer markets are also integral to the growth of small-scale sustainable viticulture. The farm-to-table movement aligns well with small-scale vineyards, where consumers value direct connections to producers through vineyard visits, wine clubs, and direct sales channels. Sustainable practices also integrate with eco-tourism, attracting visitors interested in learning about winemaking and sustainability efforts. Regulatory and industry trends further support this market. Stricter environmental regulations in major wine-producing regions push producers toward sustainable practices, while sustainability becomes a key differentiator in competitive markets. Large wineries adopting sustainable practices increase visibility and acceptance among consumers. While globalization brings standardization, the market for small-scale viticulture thrives on local identity, authenticity, and regional uniqueness.

Despite its opportunities, small-scale and sustainable viticulture faces challenges. Higher labour costs and the need for specialized knowledge can pose barriers to producers. Scaling sustainable practices to meet growing demand while maintaining quality is another difficulty. However, opportunities abound, such as collaborations with eco-conscious retailers and restaurants, creating additional revenue streams. Digital marketing and e-commerce platforms enable small-scale vineyards to reach global audiences without significant infrastructure investment.

Small-scale and sustainable viticulture represents a vital segment of the wine industry, combining environmental responsibility, cultural preservation, and premium wine production. The market is driven by increasing demand for ethical, high-quality wines and aligns with global trends toward sustainability. While challenges exist, the opportunities for small-scale producers to carve out a unique identity and appeal to discerning consumers are greater than ever. By emphasizing innovation, quality, and sustainability, small-scale viticulture is poised to play a significant role in the future of winemaking.

Small-scale and sustainable viticulture involves a complex interplay of various stakeholders, each contributing to the environmental, economic, and social sustainability of vineyard operations. This interconnected system includes vineyard owners, winemakers, agronomists, certifying organizations, researchers, policymakers, equipment providers, local communities, consumers, retailers, and environmental advocates, all of whom play distinct yet complementary roles in promoting sustainable practices.

Vineyard owners and growers are pivotal in sustainable viticulture, making critical decisions regarding farming practices, grape varieties, and vineyard design. Their focus on environmentally friendly methods—such as organic farming, biodynamic practices, and water conservation—ensures the long-term health of vineyards and surrounding ecosystems [73, 74]. Many small-scale vineyards are family-run, emphasizing quality over quantity and preserving traditional practices that foster a personal connection to the land [75]. This commitment to sustainability not only enhances the ecological integrity of the vineyards but also supports local biodiversity [76].

Winemakers are essential in converting sustainably grown grapes into high-quality wines. They collaborate closely with growers to ensure that vineyard practices align with the desired wine style and quality. In small-scale operations, winemakers often employ minimal-intervention techniques, avoiding synthetic additives and focusing on natural fermentation processes [77, 78]. This approach not only reflects a commitment to sustainability but also enhances the marketability of the wines produced, as consumers increasingly seek products that align with their values [79].

Agronomists and viticulture consultants provide vital technical support to vineyard owners and growers, offering expertise in sustainable farming practices, soil management, and pest control [80]. Their guidance helps implement innovative solutions, such as cover cropping and integrated pest management, which optimize grape production while minimizing environmental impact [81]. This collaboration between growers and consultants is worthy for the successful adoption of sustainable practices in viticulture [82].

Certifying bodies like USDA Organic and Demeter play a crucial role in sustainable viticulture by establishing standards for environmentally friendly practices and verifying compliance [83]. These certifications enhance the credibility and marketability of sustainably produced wines, providing consumers with assurance about the sustainability of their purchases [84]. The presence of such organizations fosters a culture of accountability and transparency within the industry, encouraging more producers to adopt sustainable practices [85].

Researchers and educators contribute significantly to sustainable viticulture by studying innovative farming techniques and developing disease-resistant grape varieties [86]. Their work informs best practices and provides growers with the tools and knowledge needed to adopt sustainable methods [87]. Educational institutions and agricultural extension services play a key role in disseminating this knowledge, ensuring that vineyard owners and growers are equipped to implement sustainable practices effectively [88].

Policy makers and regulators influence sustainable viticulture through laws and incentives that promote environmentally friendly practices [89]. They establish frameworks for water usage, chemical inputs, and land management, ensuring compliance with environmental standards [90]. Subsidies and grants for sustainable farming practices encourage small-scale producers to adopt these methods, thereby enhancing the overall sustainability of the viticulture sector [91].

Providers of sustainable farming equipment and technologies are critical to small-scale viticulture, offering tools that improve efficiency and reduce environmental impact [92]. Innovations such as precision irrigation systems and eco-friendly vineyard machinery enable growers to maintain profitability while adhering to sustainable practices [93]. The integration of technology in viticulture not only enhances productivity but also supports the long-term sustainability of vineyard operations [94].

Local communities are integral to the success of small-scale and sustainable viticulture, as many vineyards rely on local labour and support from nearby businesses [95]. This symbiotic relationship creates economic opportunities through employment and wine tourism, benefiting both the vineyards and the community [96]. The goodwill and participation of local residents are essential for the sustainability of vineyard operations, reinforcing the importance of community engagement in viticulture [97].

Consumers are key stakeholders in the sustainable viticulture ecosystem, driving demand for organic, biodynamic, and sustainably produced wines. Eco-conscious buyers prioritize wines with certifications and are often willing to pay a premium for products that align with their values. This growing consumer awareness reinforces the importance of transparency and sustainability in winemaking, prompting producers to adopt more environmentally friendly practices [91].

Wine retailers and sommeliers serve as intermediaries between producers and consumers, educating customers about the benefits of sustainable wines. They promote small-scale producers and curate selections that emphasize eco-friendly practices, helping to expand the market for sustainably produced wines. Their advocacy is instrumental in raising consumer awareness and fostering a culture of sustainability within the wine industry [97].

Environmental advocates and nonprofit organizations support sustainable viticulture by raising awareness about its environmental benefits and advocating for policy changes. They provide resources and training for growers and promote sustainable wine through events and educational programs. This advocacy is vital for fostering a culture of sustainability within the wine industry and encouraging more producers to adopt environmentally friendly practices.

As such, small-scale and sustainable viticulture is a collaborative effort involving a diverse network of stakeholders. Each group, from vineyard owners and winemakers to consumers and policymakers, plays a critical role in fostering environmentally responsible and economically viable practices. Together, they contribute to the preservation of traditional viticulture, the promotion of eco-friendly methods, and the production of premium-quality wines that align with modern values.

Environmental, Economic, and Social Benefits

Small-scale and sustainable viticulture practices significantly reduce the environmental impact of grape growing and winemaking by prioritizing eco-friendly methods. These practices emphasize organic farming, minimizing the use of synthetic fertilizers, pesticides, and herbicides, which protects the soil and surrounding ecosystems. Techniques like cover cropping, composting, and minimal tilling enhance soil health by preventing erosion, improving nutrient retention, and fostering biodiversity.

Research indicates that organic vineyards can substantially reduce greenhouse gas emissions and enhance soil health compared to their conventional counterparts. For instance, sustainable viticulture practices can lead to a lower environmental footprint, particularly in terms of greenhouse gas emissions associated with grape production [98]. Furthermore, studies have shown that the adoption of organic farming techniques not only mitigates climate change impacts but also improves biodiversity and soil quality, which are essential for long-term agricultural sustainability [99, 100]. The integration of life cycle assessment (LCA) methodologies in viticulture has also been highlighted as a critical tool for evaluating the environmental performance of wine production, emphasizing the importance of sustainable practices in reducing ecological footprints [101].

Water conservation is another key environmental benefit. Small-scale sustainable vineyards often use efficient irrigation systems, such as drip irrigation, or adopt dry farming techniques, significantly reducing water consumption. These practices are especially critical in regions where water scarcity poses challenges to traditional viticulture. Furthermore, by focusing on carbon-efficient processes, such as the use of renewable energy, reduced mechanization, and localized production, small-scale vineyards help mitigate their carbon footprint, contributing to broader efforts to combat climate change.

Sustainable viticulture also promotes the preservation of biodiversity within and around vineyards. By incorporating natural pest control methods, such as the use of beneficial insects and birds, these vineyards reduce chemical reliance and create habitats for diverse species. This not only supports environmental health but also fosters a balanced ecosystem that benefits grape production.

From an economic perspective, small-scale sustainable viticulture can reduce costs and enhance profitability. By minimizing reliance on synthetic inputs and large-scale machinery, small producers can lower their operating expenses. Practices like composting reduce the need for costly fertilizers, while natural pest control and disease management strategies eliminate the expense of chemical pesticides.

The economic viability of sustainable viticulture is increasingly evident, particularly in the context of the global organic wine market. According to Grand View Research, the organic wine market was valued at approximately USD 10.80 billion in 2023 and is projected to grow at a compound annual growth rate (CAGR) of 10.4% from 2024 to 2030, potentially reaching USD 21.48 billion by 2030 [102]. This growth reflects not only consumer demand for organic products but also the financial benefits for producers who adopt sustainable practices. The economic incentives for transitioning to organic viticulture are further supported by findings from Ferrara and Feo, who note that sustainable practices can lead to improved market competitiveness and consumer satisfaction [101]. Additionally, the increasing pressure from consumers and regulatory bodies for transparency regarding environmental impacts has led many producers to adopt sustainable practices as a means of enhancing their market position [103].

Small-scale sustainable producers often focus on quality rather than quantity, targeting premium and artisanal wine markets. These markets are willing to pay a premium for wines labelled as organic, biodynamic, or sustainably produced, providing producers with higher profit margins. Additionally, certifications such as USDA Organic or Demeter Biodynamic can increase marketability and attract eco-conscious consumers.

Small-scale vineyards also have the agility to experiment with innovative farming and winemaking techniques, such as permaculture or biodynamic farming, allowing them to create unique, marketable products. By participating in direct-to-consumer sales channels, such as wine clubs and farm-to-table initiatives, small-scale vineyards can further boost profitability by cutting out intermediaries and establishing personal connections with customers.

Social and ethical considerations are also driving the shift towards sustainable viticulture. A significant portion of consumers, particularly in the United States, are expressing a preference for sustainably produced wines. Reports indicate that 46% of regular wine drinkers always choose sustainable options when available, and organic wine volumes in the U.S. grew by 7% annually between 2017 and 2022, while non-organic wine volumes saw a decline [104]. This trend underscores a broader societal shift towards ethical consumption, where consumers are increasingly aware of the environmental and social implications of their purchasing decisions. The recognition of the role of sustainable practices in enhancing product quality and community well-being further supports this movement [102, 105].

Small-scale sustainable viticulture fosters community engagement and cultural preservation, offering significant social benefits. These vineyards often prioritize fair labour practices, ensuring ethical treatment and fair wages for workers. By focusing on smaller operations, they can provide better working conditions and maintain close-knit teams, contributing to local economic stability.

The preservation of traditional viticultural practices is another important social benefit. Small-scale vineyards frequently cultivate heritage grape varieties and follow age-old methods, maintaining the cultural identity of their regions. These practices are important for preserving the diversity of wine styles and the unique characteristics of regional terroirs.

Additionally, sustainable viticulture practices align with growing consumer demand for ethical and environmentally friendly products. This transparency and ethical commitment resonate with modern consumers, enhancing trust and loyalty. Small-scale vineyards also contribute to wine tourism, attracting visitors interested in learning about sustainable practices, supporting local economies, and building a connection to the land and its culture.

By integrating environmental stewardship, economic resilience, and social responsibility, small-scale and sustainable viticulture provides a comprehensive model for addressing the challenges of modern winemaking while preserving its heritage and ecological balance. This holistic approach benefits producers, consumers, and the planet alike.

The market for organic wines is not only growing in the United States but is also expanding globally. In Europe, for instance, organic wine consumption is expected to rise significantly, with projections indicating an increase from 3.7% of total wine consumption in France in 2017 to 7.7% by 2022 [104]. This expansion is indicative of a broader trend towards sustainability in the wine industry, driven by both consumer preferences and regulatory frameworks that encourage environmentally friendly practices. The increasing recognition of the importance of sustainable viticulture in maintaining the integrity of wine regions and enhancing the quality of wine products is pivotal for the future of the industry [106].

Chapter 2
The Vineyard Environment

Site Selection and Terroir

Site selection and terroir are fundamental to the success of a vineyard and the quality of the wine it produces. These factors influence the growth of grapevines, the health of the vineyard, and the characteristics of the grapes. Terroir encompasses the unique combination of climate, soil, topography, and local ecosystems that define the growing conditions of a vineyard. Together with careful site selection, terroir ensures the vineyard's potential to produce distinctive and high-quality wines.

The climate of a vineyard site dictates which grape varieties can thrive and the resulting wine's style. In cooler regions like Burgundy, France, and Central Otago, New Zealand, grapes like Pinot Noir and Chardonnay excel due to slow ripening, which preserves acidity and creates nuanced flavours. Warmer regions such as Napa Valley, USA, and Barossa Valley, Australia, are ideal for varieties like Cabernet Sauvignon and Shiraz, which develop bold, ripe flavours in abundant sunlight.

Seasonal patterns also play a significant role. For example, vineyards in Tokaj, Hungary, benefit from autumn mists that encourage noble rot, crucial for their renowned sweet wines. In Mendoza, Argentina, high-altitude vineyards experience cooler nights that preserve acidity while ripening Malbec grapes under intense sunlight. Selecting a site with the right climatic conditions ensures the balance of sugar, acidity, and flavour complexity needed for premium wine production.

Soil type and composition are fundamental to vine health and grape quality. In Bordeaux, France, gravelly soils are perfect for Cabernet Sauvignon, providing excellent drainage and forcing the vines to grow deep roots. In Mosel, Germany, slate soils retain heat, helping Riesling grapes ripen in a region with cooler temperatures. In Coonawarra, Australia, the famous terra rossa soil—a red clay over limestone—gives wines, especially Cabernet Sauvignon, their distinct minerality and structure.

Different grape varieties thrive in different soils. For example, volcanic soils in Etna, Italy, add unique minerality to wines made from Nerello Mascalese, while sandy soils in Swartland, South Africa, contribute to the elegant character of Chenin Blanc. The soil's ability to retain water, provide nutrients, and support root systems directly impacts the vine's vigour and the concentration of flavours in the grapes.

The topography of a vineyard site, including its elevation, slope, and orientation, significantly affects vine growth and grape quality. Sloped vineyards, such as those in Douro Valley, Portugal, allow for better drainage and reduce the risk of waterlogging, which can harm roots. Elevated vineyards, like those in Stellenbosch, South Africa, experience cooler temperatures, which slow ripening and preserve acidity, resulting in fresh, balanced wines.

Slope orientation is another critical factor. In the Northern Hemisphere, south-facing slopes, such as those in Piedmont, Italy, receive more sunlight, helping Nebbiolo grapes ripen fully. Conversely, in hotter regions like Hunter Valley, Australia, north-facing slopes provide some relief from intense sunlight, protecting the vines from heat stress. These factors contribute to a vineyard's ability to produce grapes with consistent ripeness and complexity.

The surrounding ecosystem plays a significant role in shaping a vineyard's microclimate and overall health. For example, in Loire Valley, France, diverse vegetation and nearby rivers create a temperate climate that supports balanced grape growth. In Okanagan Valley, Canada, the large lake moderates temperatures, preventing frost damage and extending the growing season.

Natural biodiversity benefits the vineyard by supporting pest control and reducing the need for chemical interventions. In Chianti, Italy, integrated ecosystems with forests and olive groves help maintain soil health and protect against erosion. In Yamanashi

Prefecture, Japan, terraced vineyards coexist with natural landscapes, preserving biodiversity while producing Koshu grapes known for their light and aromatic wines.

Terroir allows wines to express the unique characteristics of their growing environment, giving them a sense of place that sets them apart. In Champagne, France, the chalky soils and cool climate give sparkling wines their crisp acidity and minerality. In Marlborough, New Zealand, Sauvignon Blanc reflects the region's bright sunlight and cool nights with its vibrant tropical and citrus flavours. In Cafayate, Argentina, the combination of high-altitude sunlight and sandy soils contributes to the distinctive intensity and aromatic complexity of Torrontés.

Consumers increasingly value wines that showcase their terroir, associating them with authenticity and craftsmanship. This sense of place is particularly important in appellation-controlled regions such as Burgundy, Barolo, Italy, and Willamette Valley, USA, where terroir defines the reputation and identity of the wines.

Careful site selection also mitigates risks associated with natural hazards such as frost, drought, and excessive heat. In Alsace, France, vineyards are often planted on slopes that protect them from cold winds and frost. In Languedoc, France, windy conditions reduce humidity, lowering the risk of fungal diseases. Selecting a site with these protective features minimizes the need for costly interventions like frost fans or chemical sprays, aligning with sustainable viticulture practices.

Sustainable site selection also preserves long-term vineyard health. In regions like Colchagua Valley, Chile, where water resources are scarce, efficient irrigation methods and naturally dry climates reduce water use. In Oregon, USA, vineyards employ organic and biodynamic practices to maintain the health of their soils and ecosystems, ensuring longevity for future generations.

Site selection and terroir are foundational to vineyard success, influencing every aspect of grape growth and wine quality. The combination of climate, soil, topography, and local ecosystems defines a vineyard's ability to produce distinctive wines that reflect their place of origin. Examples from around the world—from Champagne to Marlborough and Douro Valley to Stellenbosch—illustrate how terroir shapes the character and identity of wine. Thoughtful site selection ensures not only the production of high-quality grapes but also the sustainability and resilience of the vineyard, preserving its legacy for future generations while meeting the demands of a global wine market that increasingly values authenticity and environmental stewardship.

Climate Considerations

The climate of a vineyard site is indeed a pivotal factor in determining the success of grape cultivation and the quality of the wine produced. Various climatic elements, including temperature, sunlight, rainfall, frost, and seasonal variations, play roles in grapevine development and wine quality.

Temperature is the most significant climatic factor influencing vineyard site selection. Grapevines thrive within a specific temperature range, typically between 12°C and 22°C (53.6°F to 71.6°F) [107]. In cooler climates, such as Burgundy and Mosel, grapes like Pinot Noir and Riesling benefit from slower ripening, which preserves acidity and enhances flavour complexity [107]. Conversely, warmer regions like Napa Valley and Barossa Valley support heat-tolerant varieties such as Cabernet Sauvignon and Shiraz, leading to fuller-bodied wines with higher sugar levels and bolder fruit flavours [108]. The length of the growing season is also affected by temperature; cooler regions may require early-ripening varieties to avoid frost damage, while warmer areas can support longer ripening periods for late-maturing varieties [109].

Sunlight is essential for photosynthesis, which drives vine growth and grape development. The duration and intensity of sunlight directly influence the ripening process and the accumulation of sugars in grapes [110]. Regions with longer daylight hours, such as Central Otago, allow for even ripening, while high-elevation vineyards, like those in Mendoza, benefit from intense sunlight combined with cooler temperatures, resulting in grapes with concentrated flavours and balanced acidity [111]. The orientation of vineyards also plays a role; south-facing slopes in the Northern Hemisphere maximize sunlight exposure, which is particularly beneficial in cooler climates [112].

Rainfall is also critical for vineyard health, influencing vine hydration and grape quality. Ideally, vineyards should receive adequate rainfall during the winter and spring to support growth, with minimal rain during harvest to prevent issues like berry splitting and fungal diseases [113]. Regions with low rainfall, such as Paso Robles, require irrigation to meet water needs, while wetter areas like Bordeaux necessitate proper drainage to avoid waterlogging and root rot [114]. Excessive rain during harvest can dilute grape sugars and increase disease risk, highlighting the importance of managing water resources effectively [115].

Frost presents a significant threat to vineyards, particularly in early spring when vines are budding. Late spring frosts can damage young shoots and reduce yields [116]. Regions like Champagne employ various frost mitigation strategies, including heaters and wind machines, to protect against frost damage [117]. Conversely, excessively hot conditions can lead to heat stress, causing premature ripening and negatively impacting wine quality. Strategies such as shading techniques and selecting cooler microclimates can help mitigate these risks [118].

The timing of seasonal changes is a major consideration for vineyard planning. A predictable growing season with warm springs, hot summers, and cool autumns is essential for reliable fruit set and ripening [119]. Warm, frost-free springs encourage healthy bud break, while long, warm summers allow for the development of sugars and flavour compounds. However, excessive heat can lead to overripe grapes and reduced acidity, emphasizing the need for careful management of seasonal variations [120].

Wind and humidity significantly influence vineyard microclimates and grape health. Moderate winds can reduce disease pressure by keeping vines dry, while excessive wind may damage vines or disrupt pollination [121]. High humidity levels, common in tropical regions, increase the risk of fungal diseases, necessitating effective canopy management to improve air circulation and reduce moisture retention [122].

Climate change is increasingly impacting vineyard site selection, with rising global temperatures prompting traditional wine regions to explore cooler sites at higher altitudes or latitudes [123]. For example, English wine regions are gaining recognition for sparkling wines due to warmer growing seasons, while established regions like Tuscany are adapting by employing heat-resistant grape varieties and innovative water management techniques [124]. The adaptability of vineyards to changing climatic conditions is crucial for maintaining wine quality and production sustainability [125].

The following exemplar regions and grape selections highlight how climate shapes grape choices and wine styles in each region, allowing specific varieties to thrive and reflect their unique environments and heritage.

In the USA, Napa Valley in California offers a Mediterranean climate with warm days, cool nights, and consistent sunshine. The fog from the Pacific Ocean moderates temperatures, particularly in the southern parts of the valley. This climate is ideal for Cabernet Sauvignon, Merlot, and Chardonnay. Cabernet Sauvignon thrives in the warm conditions, producing bold flavours, while Chardonnay benefits from the cooling effects of the fog, retaining its acidity and freshness. Willamette Valley in Oregon experiences a cool maritime climate with moderate rainfall and a long growing season. Summers are warm but not excessively hot, with cool evenings. Pinot

Noir, Chardonnay, and Pinot Gris are well-suited to this region, as the cooler climate allows for slow ripening, producing delicate, high-acid wines with nuanced flavours. In New York's Finger Lakes region, the cool continental climate is moderated by the lakes, extending the growing season. Riesling, Gewürztraminer, and Cabernet Franc excel here due to the vibrant acidity and minerality supported by the cool climate and water influence.

In the UK, Sussex benefits from a cool maritime climate with moderate rainfall and increasing warmth due to climate change. Summers are mild, and winters are cold but rarely extreme, making it ideal for Chardonnay, Pinot Noir, and Pinot Meunier, the foundation of sparkling wines. The chalky soils and cool conditions mirror those of Champagne, allowing for high-acidity and crisp sparkling wines. Kent has a slightly warmer and drier climate than Sussex, aided by the proximity to the English Channel. This makes it an excellent region for Bacchus, Chardonnay, and Pinot Noir. Bacchus thrives here, producing aromatic and citrusy still wines, while the other varieties are perfect for sparkling production. Essex is one of the warmest regions in the UK, with low rainfall and a long growing season. Its well-drained soils and warm climate are ideal for Pinot Noir, Bacchus, and Sauvignon Blanc, allowing these varieties to ripen fully and develop aromatic and flavourful profiles.

Australia's Barossa Valley in South Australia enjoys a warm Mediterranean climate with hot summers and cool nights. Minimal rainfall during the growing season creates perfect conditions for Shiraz, Grenache, and Cabernet Sauvignon, which produce intensely flavoured, full-bodied wines. Cool nights help preserve some acidity, balancing the boldness of these reds. Yarra Valley in Victoria has a cool to moderate maritime climate with misty mornings and warm afternoons, favouring Pinot Noir, Chardonnay, and Shiraz. The cooler conditions support elegant, high-acid wines, while Shiraz here is lighter and spicier compared to Barossa. Margaret River in Western Australia has a Mediterranean climate with mild summers and winters moderated by the Indian Ocean. This balance is ideal for Cabernet Sauvignon, Chardonnay, and Semillon, producing structured reds and fresh, elegant whites.

In Italy, Tuscany has a warm Mediterranean climate with dry summers and cool, breezy nights. Sangiovese, Vermentino, and Merlot thrive here, with Sangiovese developing bright acidity and earthy flavours in the warm days and cool nights. Coastal breezes enhance the aromatic qualities of Vermentino. Piedmont's cool continental climate, characterized by cold winters and warm, humid summers, creates ideal conditions for Nebbiolo, Barbera, and Moscato. Morning fog is a signature feature, supporting Nebbiolo's long ripening season, which develops its structured, tannic profile. Moscato benefits from the cooler temperatures for its aromatic and sweet characteristics. Sicily, with its hot Mediterranean climate and intense sunlight, is suited for Nero d'Avola, Grillo, and Catarratto. Coastal breezes moderate the heat, enabling Nero d'Avola to produce bold, ripe reds, while Grillo and Catarratto create refreshing whites.

Germany's Mosel region features a cool continental climate with steep vineyard slopes maximizing sunlight exposure. Riesling thrives in this environment, with its slate soils and mild summers supporting wines with high acidity, intense minerality, and a balance of sweetness and freshness. Rheingau's cool continental climate is moderated by the Rhine River, extending the growing season. Riesling benefits from this, developing concentrated flavours, while Spätburgunder (Pinot Noir) produces elegant, light-bodied reds. Pfalz enjoys a mild continental climate with warmer temperatures and more sunshine than Mosel or Rheingau. This supports fuller-bodied Rieslings, aromatic Gewürztraminer, and richer Pinot Noir, all of which benefit from the region's warmer and sunnier conditions.

For a new entrant into small-scale winemaking, finding reliable climate information about a region is essential for selecting grape varieties that will thrive and produce high-quality wine. Understanding factors such as local temperature ranges, rainfall patterns, sunlight exposure, frost risk, and seasonal variations ensures that the vineyard is established with a strong foundation for success. Various resources and strategies can help gather this important data.

Agricultural and viticultural research institutions often provide detailed climate information tailored to grape growing. Many regions have organizations that publish climate reports, guides, and research studies. For example, institutions like the University of California Cooperative Extension in the USA, Wine Australia, and the UK's WineGB offer resources detailing suitable grape varieties, soil types, and climatic conditions. Universities and research centres, such as Oregon State University and Geisenheim University in Germany, conduct studies on how specific grape varieties respond to regional climates. These sources are invaluable for understanding historical climate trends and making informed predictions about future conditions.

Government and meteorological services are another key resource. National agencies like the US National Weather Service or NOAA provide comprehensive data on climate and long-term weather patterns, especially for wine-producing regions in the USA. Australia's Bureau of Meteorology (BOM) offers detailed climate maps and reports critical for selecting sites in regions like Barossa or Yarra Valley. Similarly, the UK Met Office and Environment Agency provide climate data for emerging wine regions like Sussex and Kent. These services often include temperature averages, rainfall amounts, frost risks, and sunlight hours—essential details for determining grape variety viability.

Regional wine associations and growers' networks offer practical, localized insights. Associations such as Napa Valley Vintners or the Willamette Valley Wineries Association in the USA, Wine Australia, or WineGB in the UK provide collective knowledge about climate and grape performance in specific areas. These groups often organize workshops, share detailed maps, and provide consultation services, making them invaluable resources for new entrants seeking region-specific advice.

Modern technology also provides accessible tools for assessing climate conditions. Platforms like WorldClim or Climate Data Online (CDO) offer historical and predictive climate data, including temperature, precipitation, and solar radiation for regions worldwide. Terroir mapping software, such as vineyard management tools or geographical information systems (GIS), provides detailed microclimate, elevation, and soil type maps. Hyper-local weather platforms like Weather Underground or AccuWeather allow users to assess conditions specific to a plot of land, providing valuable data for site selection.

Local farmers and vineyard owners are excellent sources of firsthand knowledge about regional conditions. Building relationships with experienced growers provides practical insights into climate impacts on grape growing and the most successful varieties. Many vineyard owners are willing to share their experiences, including strategies for managing challenges like frost, drought, or pests.

Consulting with viticultural experts or agronomists can further refine climate data interpretation and grape variety recommendations. These professionals can conduct on-site evaluations and assess microclimatic factors, such as slope orientation and proximity to water bodies, which significantly affect vineyard performance. Their expertise helps tailor the site selection process to the unique conditions of the area.

Reviewing historical weather data is another critical step in site selection. Tools like WeatherSpark and national meteorological archives provide detailed data on average growing season temperatures, annual rainfall distribution, frost occurrences, and heatwave trends. Understanding these patterns helps new entrants anticipate challenges and optimize their vineyard plans.

Climate change predictions also play a significant role in modern viticulture. Organizations like the Intergovernmental Panel on Climate Change (IPCC) and regional agricultural bodies provide projections on how temperature, rainfall, and extreme weather may shift in the coming decades. This information is critical for selecting resilient grape varieties or cooler sites that may become more favourable over time, ensuring long-term sustainability.

Local terroir and microclimate analysis are equally important. Microclimates within a region can significantly impact grape growing. For example, a south-facing slope in a cool area may receive sufficient sunlight to support heat-loving grapes. Tools like soil thermometers, moisture sensors, or drone imaging can assess specific site variations in temperature, sunlight, and drainage. Proximity to moderating influences like lakes or rivers should also be factored into the decision-making process.

Finally, community workshops and training programs offered in wine regions can provide valuable educational resources. These programs often include climate and terroir education, practical tools for site evaluation, and opportunities to connect with local experts and growers.

Soil Types and Their Role in Viniculture

Soil plays a significant role in viniculture, influencing vine health, grape quality, and the characteristics of the resulting wine. The composition and type of soil directly affect various aspects of grapevine growth, including nutrient availability, water retention, drainage, and temperature regulation. Understanding these relationships is essential for winemakers to select appropriate vineyard sites and manage them effectively to produce high-quality wines.

The composition of soil, which includes minerals, organic matter, water, and air, is fundamental to grapevine health. The three primary components—sand, silt, and clay—determine soil texture and influence root interactions with the soil. Sandy soils provide excellent drainage and aeration but have low nutrient and water retention capabilities, which can limit vine vigour [126]. In contrast, clay soils retain significant moisture and nutrients, making them suitable for grape varieties that require consistent water supply, such as Merlot and Syrah [127]. Loamy soils, which contain a balanced mix of sand, silt, and clay, are often considered ideal for viticulture due to their moderate water retention and good drainage properties [128].

Different soil types impart unique qualities to grapes, affecting vine vigour, grape ripeness, and wine characteristics. For instance, gravelly soils, like those found in Bordeaux, offer excellent drainage and heat reflection, which aids in the even ripening of grapes such as Cabernet Sauvignon, resulting in structured wines with bold flavours [129]. Limestone and chalk soils, prevalent in regions like Champagne and Burgundy, retain moisture while providing good drainage, supporting high-acid grapes like Chardonnay and Pinot Noir, which yield wines with bright acidity and minerality [130]. Conversely, clay soils, found in regions like Pomerol, contribute to full-bodied, concentrated wines due to their moisture retention capabilities [131].

Slate soils, common in the Mosel region, reflect heat and help ripen grapes in cooler climates, benefiting varieties like Riesling [132]. Volcanic soils, such as those in Etna, are rich in minerals and drain well, contributing to wines with distinctive minerality and complexity [133]. Sandy soils, known for their excellent drainage, reduce vine vigour and pest pressure, allowing for concentrated fruit, as seen in regions like Ribera del Duero and Stellenbosch [134].

Drainage and water retention are critical characteristics in viticulture. Soils that drain well, such as gravel and sand, are preferred by grapevines, which thrive in relatively dry conditions. Excessive water retention can lead to root rot and overly vigorous vine growth, diluting grape flavours [135]. In regions with high rainfall, like Bordeaux, gravel soils prevent waterlogging, ensuring vine health and consistent fruit quality [136]. Conversely, in arid regions such as Mendoza, soils with moderate water retention, like loam or clay, are essential for sustaining vines without frequent irrigation [137].

Grapevines do not require highly fertile soils; in fact, excessive nutrients can lead to overly vigorous growth at the expense of fruit quality. Soils with moderate fertility are ideal, encouraging balanced vine growth. Key nutrients include nitrogen, potassium, and phosphorus, along with trace elements like magnesium and iron [138]. Volcanic soils often provide rich mineral content that enhances vine health and contributes to unique wine flavours, while sandy soils, being less fertile, naturally limit vine vigour and promote concentrated fruit [139].

Soil temperature significantly influences the ripening process by affecting root activity and nutrient uptake. Light-coloured soils, such as chalk or limestone, reflect sunlight and help keep vineyards cooler, which is beneficial in hot climates. Dark soils, like slate, absorb heat, aiding ripening in cooler regions [140]. These variations allow winemakers to select grape varieties that align with the thermal properties of the soil, optimizing growth conditions [141].

Soil is a fundamental component of terroir, the concept that the environment shapes the character of wine. The interplay between soil, climate, and geography creates unique growing conditions for each vineyard. For example, the combination of slate soils and a cool climate in the Mosel region produces Rieslings with remarkable minerality and precision, while the volcanic soils of Mount Etna yield wines with intense flavours and complexity [142]. Understanding these interactions enables winemakers to align soil types with grape varieties, achieving optimal growth conditions and producing exceptional wines that reflect their origins [143].

Determining soil type and quality is a critical step in establishing a vineyard, as the soil significantly influences vine health, grape quality, and the overall success of the vineyard. Understanding the physical, chemical, and biological properties of the soil enables winemakers to select the most suitable grape varieties and implement effective management practices to maximize productivity and quality.

The first step in evaluating soil is collecting representative soil samples from the potential vineyard site. Samples should be taken from multiple depths, such as 0–30 cm and 30–60 cm, to assess the characteristics of the root zone. Sampling should also cover various areas of the site, especially where soil properties appear to vary, such as on slopes, flat areas, or wet spots. Using tools like soil augers or shovels ensures samples remain uncontaminated and provide an accurate representation of the site's soil profile.

Visual and physical assessments of the soil provide immediate insights into its type and texture. Rubbing soil between the fingers helps determine its composition—sand feels gritty, silt feels smooth, and clay feels sticky. Soil colour can also indicate organic content and drainage capabilities, with darker soils often being richer in organic matter. Observing soil structure and how it aggregates offers clues about aeration and root penetration. Drainage properties can be gauged by watching how water behaves in the soil, with poorly draining soils often having high clay content and sandy soils draining too quickly.

Laboratory soil testing offers detailed insights into soil quality. Testing includes analysing pH levels, which affect nutrient availability, as grapevines generally thrive in a pH range of 6.0 to 7.5. Nutrient content testing focuses on key elements like nitrogen, phosphorus, potassium, magnesium, and calcium, all critical for vine growth. Organic matter levels, which influence fertility and water-holding capacity, are also measured, as well as soil salinity, since high salt content can damage grapevines. Laboratory testing provides precise data on the soil's chemical composition and highlights any deficiencies that may need correction.

Assessing soil drainage is another necessary step. Poor drainage can lead to root rot, while overly dry soils may require irrigation. A simple percolation test, where a hole is filled with water to observe its drainage rate, can identify soil tendencies. Water table depth should also be checked, especially in low-lying areas, as high water tables can saturate roots and limit oxygen availability. These insights help determine whether drainage systems or irrigation infrastructure may be needed.

Soil depth directly affects vine rooting potential. Deep soils, typically 1–2 meters, allow roots to access water and nutrients during dry periods, supporting consistent vine growth. Shallow soils require careful management to prevent water stress, especially during the growing season.

Analysing soil structure and compaction is essential for understanding how soil supports root growth and water movement. Using a penetrometer to measure compaction levels can identify areas where soil restricts root penetration and water infiltration. Evaluating the soil's friability, or its ability to crumble easily, provides additional information about how well it supports healthy root systems.

Soil biology also plays a vital role in vineyard health. Microbial activity, including the presence of bacteria, fungi, and earthworms, improves nutrient availability and enhances soil structure. Biological testing, offered by some laboratories, provides insights into microbial populations and can guide decisions on practices like composting, cover cropping, or adding organic amendments to boost soil life.

Advanced tools like Geographic Information Systems (GIS) and soil mapping software help analyse the soil characteristics across a site. These tools provide detailed information about topography, elevation, slope, and soil variability, which are all factors influencing drainage, temperature regulation, and sunlight exposure. Mapping soil variations can identify areas requiring different management practices, allowing for optimized vineyard layout and resource allocation.

Soil evaluation must also consider the region's climate and microclimate. Soil water-holding capacity should align with rainfall patterns and irrigation potential. In cooler climates, soils that retain or reflect heat, such as slate or gravel, can enhance ripening by moderating temperature fluctuations. These considerations ensure the soil supports grape varieties best suited to the site's environmental conditions.

Consulting with experts like soil scientists, agronomists, or viticultural consultants provides professional evaluations and tailored recommendations. These experts can suggest soil improvement techniques, such as adding amendments, installing drainage systems, or implementing cover crops, to enhance soil quality and vineyard performance.

Grapevine Structures

The grapevine is a perennial plant with a complex structure, each part playing a vital role in supporting the vine's growth, fruit production, and overall health. Understanding the anatomy of the grapevine is essential for effective vineyard management. It enables growers to optimize practices such as pruning, training, and disease prevention, ensuring high-quality grape production.

The root system anchors the grapevine to the soil and serves as its primary source of water and nutrients. Grapevine roots can extend several meters deep and laterally, depending on soil type, climate, and irrigation practices. Roots perform three main functions: anchorage, absorption, and storage. They stabilize the vine, enabling it to withstand environmental stresses like wind and soil erosion. They also absorb water and essential nutrients, such as nitrogen, potassium, and phosphorus, critical for vine growth and fruit development. During dormancy, the roots act as reservoirs for carbohydrates and other nutrients, ensuring the vine has sufficient energy to resume growth in spring. The depth and spread of roots influence the vine's resilience and productivity, varying based on soil type and vineyard management practices.

The trunk is the central, woody structure of the grapevine that connects the root system to the above-ground parts. It acts as a conduit for transporting water, nutrients, and sugars between the roots and the canopy. Over time, the trunk thickens and develops bark, providing protection against physical damage and environmental stress. The trunk's height and shape depend on the vineyard's training system. In small-scale operations, the trunk is often pruned and trained to specific heights to facilitate canopy management and harvesting.

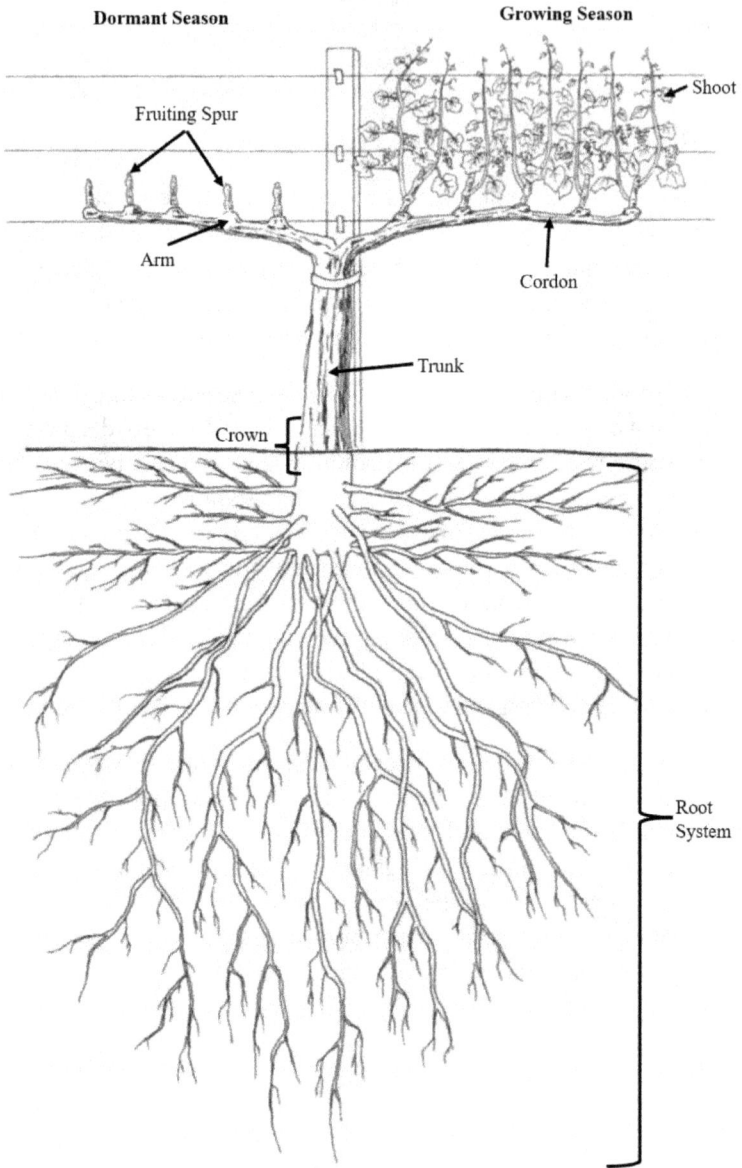

Figure 8: Grapevine structure.

Extending from the trunk are the arms and cordons. Arms are short, woody branches, while cordons are longer, horizontal extensions found in certain training systems, such as the spur-pruned cordon system. These structures serve as the foundation for annual shoot and fruit production. They support spurs or canes, which bear the new shoots and grape clusters each season. Regular pruning and training of arms and cordons ensure balanced vine growth, prevent overcrowding, and optimize fruit yield and quality.

Shoots represent the annual, green growth that emerges from buds on spurs or canes in spring. They play a critical role during the growing season by supporting leaves, flowers, tendrils, and grape clusters. Leaves on the shoots capture sunlight, producing the energy required for vine growth and grape ripening through photosynthesis. Shoots also provide a framework for the vine's reproductive structures. At the end of the growing season, shoots lignify (become woody) and form canes, which are pruned back during winter.

Nodes and internodes are structural elements of the shoots and canes. Nodes are small, swollen areas along the shoot or cane where leaves, tendrils, and clusters develop. Each node contains a bud that can grow into a shoot, leaf, tendril, or flower cluster. The spacing between nodes, or internodes, reflects vine vigour, light exposure, and nutrient availability. Longer internodes indicate vigorous growth, while shorter ones may signal stress or poor growing conditions.

Buds are dormant growth points located at the nodes, categorized as primary, secondary, or tertiary. Primary buds are the most productive, giving rise to shoots that bear fruit. Secondary buds serve as backups, producing limited or no fruit if the primary buds are damaged. Tertiary buds rarely grow unless both primary and secondary buds fail. The health and development of buds are key for the following season's productivity, making bud management a priority in vineyard practices.

Leaves are vital for photosynthesis, the process that provides the energy necessary for the vine's growth and fruit development. Leaves consist of a broad blade (lamina) and a stalk (petiole) that attaches to the shoot. Besides photosynthesis, leaves regulate water loss through transpiration and facilitate gas exchange via stomata. Proper leaf positioning ensures optimal sunlight exposure and air circulation, reducing disease risks and promoting fruit ripening.

Tendrils are thin, coiling structures that help the grapevine climb and secure itself to supports like trellises or neighbouring plants. Located opposite the leaves on the shoots, tendrils play a purely structural role and do not contribute to photosynthesis or fruit production.

Flowers and clusters mark the reproductive stage of the grapevine. The flowers are small, green, and self-pollinating, growing in clusters that eventually develop into grape bunches. The size and compactness of fruit clusters depend on the grape variety and environmental conditions. Successful flowering, monitored closely by growers, is critical as weather conditions during this stage can significantly impact yield.

Grape berries, the end product of the vine's growth cycle, consist of three main parts: skin, flesh, and seeds. The skin contains pigments, tannins, and aromatic compounds, influencing the wine's colour, structure, and flavour. The flesh holds water, sugars, and acids that contribute to the wine's body and balance. Seeds contain tannins and oils that can impact the wine's flavour profile if extracted during winemaking. The composition of the grape berry evolves throughout the ripening process, directly affecting the wine's final quality.

Each structure of the grapevine plays an integral role in its growth, reproduction, and ability to produce high-quality grapes. By understanding and managing these structures, vineyard owners can enhance vine health, maximize yields, and produce wines that reflect the unique characteristics of their terroir.

The Grapevine Growth Cycle

Typically, the grapevine growth cycle is an annual process that consists of distinct stages, each critical for the development of the vine and the production of high-quality grapes. The cycle begins in early spring and ends with dormancy in winter, driven by environmental conditions such as temperature, sunlight, and rainfall. Understanding these stages helps vineyard managers optimize care and interventions for healthy vine growth and successful wine production.

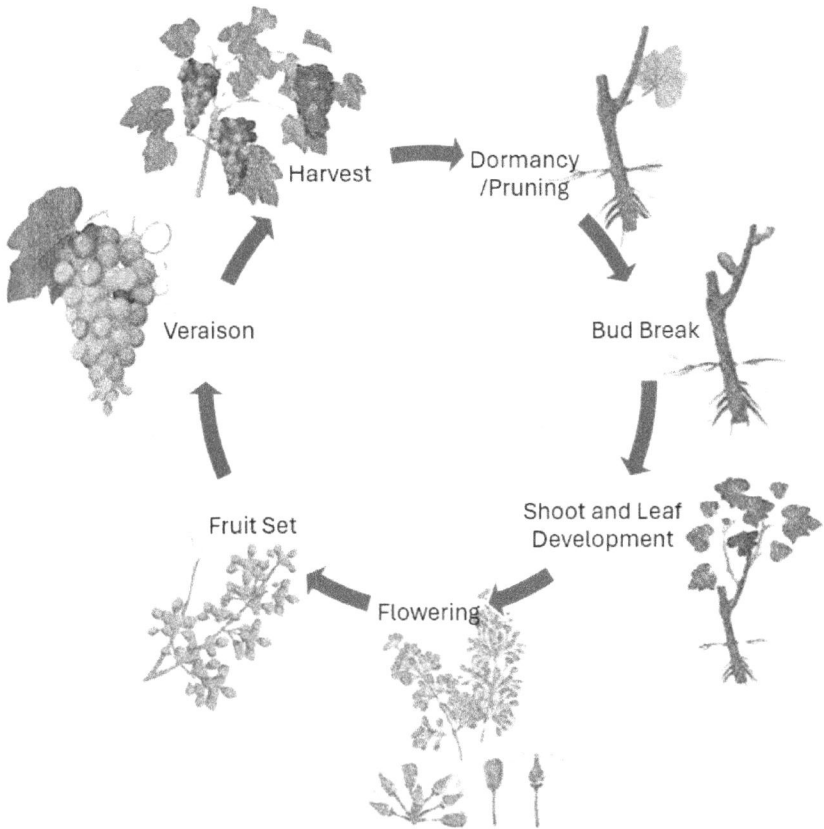

Figure 9: The Grapevine Growth Cycle.

1. Dormancy (Winter)

Dormancy occurs during the cold winter months when the vine rests. At this stage, the vine appears lifeless, with no visible growth above the soil, but vital processes continue underground. The roots remain active, storing energy and nutrients in preparation for the next growth season.

- **Key Activities**:
  - The vineyard is pruned to shape the vine for optimal growth in the coming year.

- o Weeds are managed, and soil health is monitored to prepare for spring.

Dormancy ends as temperatures rise in early spring.

Figure 10: Dormant Grape Vines, Old Mission Peninsula. Joe Ross, CC BY-SA 2.0, via Flickr.

2. Bud Break (Early Spring)

As the weather warms and soil temperatures rise above 10°C (50°F), dormant buds on the vine start to swell and burst open. This marks the beginning of visible vine activity. Small shoots and leaves emerge, initiating photosynthesis, which fuels vine growth.

- **Key Factors**:
    - Frost protection may be required in cooler climates to safeguard tender shoots.
    - Nutrient and water availability influence early growth.

Bud break sets the stage for the vine's development and is a critical time for monitoring vineyard health.

3. Shoot and Leaf Development (Mid-Spring)

Following bud break, shoots grow rapidly, and leaves expand to form the vine's canopy. This canopy is essential for photosynthesis, which provides energy for growth and development. At this stage, flower clusters begin to form but are not yet open.

- **Key Activities**:
    - Canopy management begins, including training and positioning shoots to maximize sunlight exposure and airflow.
    - Pest and disease control measures are implemented as the warmer, wetter weather can encourage problems like powdery mildew.

Shoot and leaf development is vital for building the vine's energy reserves and supporting future fruit development.

4. Flowering (Late Spring to Early Summer)

Flowering begins as tiny flowers open in clusters, marking the reproductive stage of the vine. This process typically lasts 1–2 weeks and is highly sensitive to weather conditions.

- **Key Processes**:

  o Pollination occurs, often facilitated by wind. Most grapevines are self-pollinating, meaning they do not rely on insects.

  o Warm, dry weather promotes successful flowering, while rain or wind can disrupt the process and reduce fruit set.

The success of flowering directly impacts the number of grapes that will develop, influencing yield.

Figure 11: Grapevines during flowering. Andy Melton, CC BY-SA 2.0, via Wikimedia Commons.

5. Fruit Set (Early Summer)

Fruit set follows flowering when fertilized flowers develop into small green berries. At this stage, the berries begin to grow, but they are hard, green, and highly acidic.

- **Key Factors**:
    - Weather during this period affects fruit set; hot or wet conditions can lead to uneven berry development.
    - Balanced vine nutrition ensures healthy fruit set and cluster formation.

Fruit set determines the vineyard's potential yield and is critical for maintaining quality.

6. Veraison (Mid-Summer)

Veraison marks the onset of grape ripening, as the berries undergo physical and chemical changes.

- **Key Changes**:
    - For red varieties, the berries change colour from green to red or purple. White varieties turn golden or translucent.
    - Sugars accumulate, and acidity begins to decrease.
    - Phenolic compounds, such as tannins, develop, contributing to the wine's structure and flavour.

Veraison is a time of transformation, and water stress or excessive canopy growth is carefully managed to promote ripening.

7. Ripening (Late Summer to Early Fall)

During ripening, berries continue to soften, accumulate sugars, and develop flavours. Acidity decreases, while aroma compounds intensify. This stage determines the grape's final quality.

- **Key Activities**:

- The vineyard is closely monitored for ripeness indicators, such as sugar levels (measured as °Brix), acidity, and tannin maturity.

- Irrigation may be reduced or stopped to concentrate flavours and prevent dilution.

- Netting or other protective measures are often used to safeguard grapes from birds or pests.

Optimal timing of harvest is vital to achieving the desired wine style and quality.

8. Harvest (Fall)

Harvest is the culmination of the grapevine growth cycle when ripe grapes are picked. Timing is critical and depends on the desired wine style, grape variety, and environmental conditions.

- **Manual vs. Mechanical Harvesting**:

  - Small-scale operations often prefer hand harvesting for precision and care.

  - Larger vineyards may use mechanical harvesters for efficiency.

Harvest timing affects wine characteristics, such as sweetness, acidity, and tannin balance.

9. Post-Harvest and Leaf Fall (Late Fall)

After harvest, the vine redirects its energy to replenish root reserves for the next cycle. Leaves begin to change colour and fall as the vine prepares for dormancy.

- **Key Activities**:

  - Compost or mulch may be added to the soil to restore nutrients.

  - Cover crops may be planted to improve soil health and prevent erosion during winter.

This transitional phase ensures the vine remains healthy and resilient for the next growth season.

10. Dormancy (Winter)

The cycle ends with dormancy, as the vine enters a state of rest. The roots continue to store energy and nutrients, completing the annual cycle and preparing for a new season.

Water Availability and Management

Water availability and management are critical factors influencing viticulture, affecting vine health, grape quality, yield, and the sustainability of vineyards. Grapevines require a precise balance of water throughout their growth cycle, as both excess and deficiency can lead to significant challenges. Effective water management strategies are essential for optimizing grape development, ensuring high-quality wine production, and adapting to environmental conditions, including climate change and regional water scarcity.

Water is fundamental to the physiological processes of grapevines, including photosynthesis, nutrient transport, and cell expansion. It supports vegetative growth during the early growing season and is imperative for berry formation and ripening. The availability of water directly influences the size, sugar content, acidity, and phenolic compounds in grapes, which are vital determinants of wine quality [117, 144]. Inadequate water can induce water stress, causing vines to prioritize survival over fruit development, resulting in smaller berries, reduced yields, and unbalanced grapes with high acidity and low sugar levels [145, 146]. Conversely, excessive water can lead to overly vigorous vine growth, diluted fruit flavours, and increased disease susceptibility due to excess canopy shading and humidity [147, 148].

The water needs of grapevines vary throughout their growth cycle, with specific requirements at different stages. From bud break to flowering, adequate water is crucial for early vine growth and the development of flower clusters; deficits during this stage can reduce flowering and fruit set [149, 150]. During the fruit set to veraison stage, moderate water supply supports berry growth and prevents excessive vegetative growth, ensuring energy is directed toward fruit development [151, 152]. Controlled water stress during the veraison to harvest period can enhance the concentration of sugars and phenolic compounds while maintaining acidity, whereas excessive water can dilute grape flavours and increase disease risks [117, 153]. Minimal water is needed during dormancy, although some soil moisture is necessary to maintain root health [154, 155].

Water availability varies significantly across wine-growing regions, influencing viticultural practices and grape varieties. In arid regions such as Mendoza (Argentina) and Barossa Valley (Australia), limited rainfall necessitates irrigation to ensure a consistent water supply, with grape varieties like Malbec and Shiraz thriving under controlled irrigation [152, 156]. Conversely, rainfed regions like Bordeaux (France) and the Willamette Valley (USA) often rely on natural rainfall, although variations in annual rainfall may require supplemental irrigation or drainage solutions [149, 157]. In cool and wet regions, such as Mosel (Germany) and parts of New Zealand, excessive rainfall can lead to waterlogging and root rot, necessitating effective soil drainage and canopy management to maintain vine health [144, 158].

Water scarcity can severely limit vine growth, reduce yields, and impact grape quality. Prolonged droughts may excessively stress vines, leading to unbalanced grapes with low sugar and high tannin levels. In extreme cases, drought conditions can threaten the long-term viability of vineyards by weakening root systems and depleting soil moisture reserves [145, 146]. Conversely, excessive water can saturate soils, reducing oxygen availability to roots and causing root rot. Overly vigorous vegetative growth from abundant water can lead to dense canopies, increasing humidity and disease risks such as powdery mildew and botrytis [147, 148]. Wet conditions during ripening or harvest can dilute grape flavours and disrupt sugar accumulation, making berries more susceptible to splitting [153, 155].

Effective water management is essential for balancing vine water needs with environmental conditions. Key strategies include the use of irrigation systems, such as drip irrigation, which delivers precise amounts of water to the root zone, minimizing waste [152, 159]. Dry farming techniques, which rely on natural precipitation, encourage deep root growth and produce concentrated, high-quality fruit but require careful site selection and soil management [148, 156]. In wet regions, drainage systems help prevent waterlogging and maintain optimal soil moisture levels [149, 157]. Additional strategies include mulching to retain soil moisture, planting cover crops to improve soil structure and water infiltration, and employing regulated deficit irrigation to induce mild stress that enhances flavour concentration [146, 160].

Climate change presents new challenges for water management in viticulture, as rising temperatures and shifting rainfall patterns may exacerbate water scarcity in some regions while increasing flood risks in others [112, 155]. Adapting to these changes involves implementing water-efficient technologies, selecting drought-resistant grape varieties, and improving vineyard design to optimize water use [154, 161]. Innovations such as soil moisture sensors, weather forecasting systems, and satellite imagery can help monitor water needs and predict irrigation requirements, enabling precise management [152, 159]. Regions facing chronic water shortages may explore alternative water sources, such as recycled wastewater, to sustain viticulture without depleting natural resources [112, 160].

Ideal Vineyard Environment for Small-Scale Sustainable Wine Growing and Making

Creating a vineyard environment suitable for small-scale sustainable wine growing and making requires thoughtful planning and the integration of ecological balance, minimal environmental impact, and the ability to produce high-quality grapes. A sustainable vineyard must harmonize with its natural surroundings while fostering optimal conditions for vine growth and wine production.

A vineyard's climate is one of the most critical factors influencing vine health and grape quality. For small-scale sustainable operations, selecting a region with a climate suited to the desired grape varieties minimizes the need for artificial interventions such as irrigation or frost protection. Grapevines thrive in areas where annual temperatures range between 10°C and 20°C (50°F to 68°F). Cooler climates are ideal for high-acid varieties like Pinot Noir and Chardonnay, while warmer climates are better suited for fuller-bodied varieties like Cabernet Sauvignon and Shiraz. Moderate, well-timed rainfall is also important, with rainfall in winter and spring reducing the need for irrigation. A defined growing season with warm summers and cool nights supports balanced grape development by allowing sugars, acids, and phenolic compounds to mature harmoniously.

Healthy soil is the foundation of sustainable viticulture, providing balanced vine growth, essential nutrients, and efficient water management. Well-draining soils, such as gravel, sandy loam, or limestone-rich soils, prevent waterlogging and support deep root growth. Soil with moderate fertility encourages controlled vine vigour, while high levels of organic matter improve soil structure, water retention, and nutrient availability, reducing the need for synthetic fertilizers. A biologically active soil with diverse microbial life supports nutrient cycling and suppresses diseases, which can be enhanced through practices like composting and cover cropping.

Water management is another vital component of sustainable vineyards, particularly for small-scale operations aiming to reduce environmental impact and ensure long-term viability. Drip irrigation systems are highly efficient, delivering precise amounts of water to the roots when required. In regions with reliable rainfall, dry farming is preferable. Proper drainage systems, especially on sloped terrains, prevent erosion and waterlogging, while rainwater harvesting and recycling systems further support sustainable practices.

Biodiversity plays a significant role in enhancing vineyard sustainability. Planting cover crops such as legumes or grasses between vine rows improves soil health, prevents erosion, and attracts beneficial insects. Hedgerows and buffer zones of native vegetation provide habitats for pollinators and natural pest predators, reducing the need for chemical inputs. Encouraging wildlife through birdhouses or insect hotels promotes ecological balance and further minimizes the use of synthetic pesticides.

A well-designed vineyard layout optimizes natural resources and minimizes environmental impact. Slope orientation is critical, with south-facing slopes in the Northern Hemisphere (or north-facing slopes in the Southern Hemisphere) maximizing sunlight exposure, especially in cooler climates. Adequate row spacing and trellising systems ensure good air circulation, reducing the risk of humidity-related diseases like mildew. On steep slopes, terraces prevent soil erosion and improve water retention.

Minimizing chemical use is a cornerstone of sustainable viticulture. Organic practices that replace synthetic fertilizers, herbicides, and pesticides with natural alternatives like compost, mulch, and biological pest control protect both soil health and the surrounding ecosystem. Integrated Pest Management (IPM) further reduces chemical use by monitoring pest levels and applying targeted, environmentally friendly solutions.

Renewable energy and resource efficiency are key to the environmental and economic sustainability of small-scale vineyards. Solar panels can power irrigation systems, cellar equipment, and other operations, while energy-efficient machinery reduces operational costs and environmental impact. Recycling waste, such as composting grape pomace for use as organic fertilizer, completes the nutrient cycle and minimizes waste generation.

Social and economic sustainability are equally important in creating a viable vineyard environment. Fair labour practices ensure safe working conditions and fair wages for vineyard workers, fostering a socially responsible operation. Collaborating with local suppliers, winemakers, and markets strengthens community ties and promotes the vineyard's identity. Direct-to-consumer sales, such as wine clubs or tasting room visits, reduce the carbon footprint associated with distribution while providing a personalized connection with customers.

Adapting to climate change is essential for long-term sustainability. Planting drought-resistant grape varieties, such as Grenache in arid regions or Riesling in cooler climates, ensures consistent yields under changing conditions. Agroforestry techniques, such as integrating trees or shrubs into the vineyard, provide shade, reduce water evaporation, and enhance biodiversity. Practices like no-till farming and cover cropping sequester carbon in the soil, mitigating the vineyard's contribution to climate change.

An ideal environment for small-scale sustainable wine growing and making combines favourable climate, healthy soil, efficient water management, and biodiversity integration. By minimizing chemical use, optimizing natural resources, and incorporating renewable energy, small-scale producers can create high-quality wines while preserving the environment and supporting their local community. This holistic approach not only ensures economic viability but also strengthens the vineyard's identity and its connection to the unique characteristics of its terroir.

Assessing Microclimates

Microclimates are critical for the success of small-scale wine growing, as they create localized environmental conditions that significantly influence vine health, grape quality, and the unique characteristics of the resulting wine. A microclimate is defined by specific variations in temperature, humidity, sunlight, wind, and precipitation within a vineyard, which can differ markedly from the broader regional climate. Understanding these localized conditions allows small-scale wine growers to optimize their practices for better outcomes.

Microclimates are shaped by various geographic, topographic, and environmental factors. Topography, including elevation, slope, and aspect, plays a vital role in determining sunlight exposure and wind patterns, which can lead to significant temperature variations within a vineyard [162, 163]. Proximity to water bodies, such as lakes or rivers, can moderate temperatures, reducing risks associated with frost and extreme heat [162, 164]. Soil properties, including type, colour, and drainage capacity, also affect how heat and moisture are retained or reflected, further influencing vine health [163, 164]. Additionally, surrounding vegetation can alter wind flow and shade, creating distinct microclimate zones within a vineyard [162, 163].

By analysing these factors, growers can identify areas within their vineyard that offer unique growing conditions. For instance, studies have shown that the microclimate can significantly affect the composition of grape berries, including the concentration of flavonoids and other important compounds [165, 166]. This understanding is fundamental for making informed decisions regarding vineyard management and grape variety selection.

Microclimates directly impact grapevine growth, fruit ripening, and overall wine quality. For small-scale wine growers, leveraging microclimates is essential for several reasons. First, selecting grape varieties that match specific microclimates ensures optimal ripening and flavour development [167, 168]. For example, the composition of grapes can vary significantly based on the microclimate, affecting the wine's sensory characteristics [169, 170].

Moreover, microclimate variations enhance terroir expression, contributing to the unique characteristics of the wine that reflect its specific location [164, 170]. Understanding localized climatic conditions allows growers to mitigate risks associated with adverse weather, such as frost pockets or excessive heat zones, enabling them to implement targeted interventions [163, 164]. For instance, practices like early leaf removal can modify the fruit zone microclimate, enhancing sunlight exposure and improving grape quality [168, 171].

Methods for Assessing Microclimates

To evaluate microclimates for small-scale wine growing, growers employ a combination of observational techniques, tools, and data analysis to identify localized variations within the vineyard that affect vine growth and grape quality. These methods help tailor vineyard management practices to optimize grape production and ensure sustainable operations.

Temperature monitoring is a key aspect of microclimate assessment. Installing weather stations or temperature sensors at various points within the vineyard allows growers to track daily and seasonal temperature variations. This data is central for identifying frost-prone areas, zones of heat accumulation during the growing season, and diurnal temperature swings (the difference between daytime and nighttime temperatures). These variations can significantly influence vine health and grape ripening.

Understanding sunlight and solar radiation is another essential component. Sunlight exposure can be analysed by observing slope orientation (aspect) and the impact of nearby obstructions, such as trees or buildings. Light meters and solar mapping tools are used to measure the intensity and duration of sunlight in specific vineyard areas. Adequate sunlight promotes photosynthesis and ripening, while shaded areas may delay fruit development. This information helps growers determine where to plant different grape varieties and how to arrange vineyard rows for maximum exposure.

Wind patterns are also crucial to microclimate assessment. Monitoring wind strength and direction identifies areas prone to strong winds or zones with limited air circulation. Moderate winds improve airflow, reducing humidity and the risk of fungal diseases, while excessive winds can damage vines or cause moisture loss. Understanding wind patterns allows growers to implement windbreaks or adjust trellising systems to protect the vines and optimize airflow.

Moisture levels and humidity are measured using soil moisture sensors and hygrometers, which provide insights into both soil and air moisture content. High humidity can increase the risk of fungal diseases like mildew, while excessively dry areas may require irrigation to maintain vine health. Knowing the moisture distribution across the vineyard helps growers implement targeted irrigation strategies or drainage systems to ensure consistent growing conditions.

Assessing soil temperature and composition is equally important. Soil temperature variations are recorded across the vineyard, as darker soils absorb and retain more heat than lighter soils. These variations influence root activity and early vine growth, particularly in cooler climates. Soil composition also affects heat retention, drainage, and nutrient availability, all of which are critical for vine development.

Lastly, mapping variations within the vineyard provides a comprehensive view of its microclimates. Tools like Geographic Information Systems (GIS) and drone imaging create detailed maps of topography, slope, and aspect. These maps help growers divide the vineyard into zones based on variations in temperature, sunlight, and wind, allowing for tailored management practices. For example, growers can assign grape varieties to zones best suited to their specific growing requirements or implement precise canopy management strategies.

By combining these methods, small-scale wine growers can develop a detailed understanding of their vineyard's microclimates, enabling them to optimize vine health, grape quality, and overall productivity while maintaining sustainability.

Applications of Microclimate Assessment

On-site measurements are indispensable for understanding and managing the microclimate of a vineyard. Tools such as thermometers, soil sensors, light meters, and anemometers provide real-time data on temperature, soil conditions, sunlight intensity, and wind patterns. These measurements allow growers to monitor environmental conditions closely and make informed decisions about vineyard management practices, such as irrigation, canopy management, and frost protection.

Satellite and drone imaging have revolutionized vineyard mapping by offering precise, high-resolution visuals of vineyard topography and variability. These technologies can identify subtle differences in slope, elevation, and aspect that influence microclimate conditions. By analysing this data, growers can create detailed maps to optimize vineyard layout, zone management, and resource allocation, ensuring that each section of the vineyard is managed according to its unique conditions.

Collaborating with viticultural consultants or agricultural extension services provides additional expertise for advanced microclimate analysis and tailored recommendations. These professionals bring experience and scientific knowledge to help growers interpret data and implement effective strategies. From advising on grape variety selection to suggesting innovative practices, their guidance enhances the overall success and sustainability of small-scale vineyards.

For small-scale wine growers, assessing microclimates allows for precise and efficient use of resources, improving both grape quality and vineyard sustainability. Tailoring vineyard practices to microclimate conditions maximizes the potential of limited land and ensures that each section of the vineyard contributes to the overall success of the operation. Additionally, the distinct characteristics of microclimate-influenced wines can provide a competitive edge in the market by highlighting their unique terroir.

Biodiversity and Ecosystem Health

Biodiversity and ecosystem health are important components of sustainable viticulture, influencing not only the ecological balance within vineyards but also the economic viability of wine production. By fostering diverse biological communities, vineyards can enhance resilience against environmental stressors, reduce reliance on chemical inputs, and contribute to the conservation of local ecosystems.

Biodiversity in vineyards includes a variety of organisms such as plants, animals, fungi, and microorganisms, which interact to maintain ecosystem balance. One of the primary benefits of biodiversity is natural pest control. Diverse ecosystems attract beneficial insects and birds that manage pest populations effectively. For instance, ladybugs and lacewings prey on aphids, while birds can help control caterpillar populations that threaten vines [172, 173]. This natural pest management reduces the need for chemical pesticides, aligning with sustainable agricultural practices.

Moreover, healthy ecosystems support soil microbial communities that play a vital role in disease prevention. Beneficial microbes can suppress pathogens and enhance vine immunity, thus reducing the incidence of diseases like downy mildew and botrytis [174]. The presence of diverse soil organisms contributes to nutrient cycling and organic matter breakdown, which are essential for maintaining soil health and fertility [174].

Biodiversity also enhances climate resilience in vineyards. Diverse plant cover, including cover crops and hedgerows, can mitigate the effects of extreme weather events, such as heavy rains and droughts, by improving soil structure and moisture retention [173]. This resilience is critical as climate change continues to pose challenges to agricultural systems globally.

To promote biodiversity, several practices can be implemented in vineyards. The use of cover crops, such as legumes and wildflowers, not only increases biodiversity but also improves soil health by preventing erosion and enhancing water infiltration [174]. Additionally, hedgerows and buffer zones can provide habitats for wildlife, further supporting ecosystem health and reducing chemical runoff into surrounding areas [173].

Integrated Pest Management (IPM) is another effective strategy that combines biological, cultural, and mechanical methods to control pests while minimizing chemical use. For example, creating habitats for natural predators and using pheromone traps can significantly reduce pest populations without relying heavily on synthetic pesticides [172, 173]. Furthermore, maintaining wildlife corridors within or adjacent to vineyards can enhance genetic diversity among species and prevent habitat fragmentation, which is critical for maintaining healthy ecosystems [173].

Organic and biodynamic practices also contribute to biodiversity by avoiding synthetic inputs and fostering natural ecological processes. These methods not only support the health of the vineyard ecosystem but also appeal to a growing consumer base that values environmentally friendly production methods [175, 176].

Several regions have successfully implemented biodiversity-friendly practices in their vineyards. In Champagne, France, vineyards have adopted hedgerow planting and cover cropping, which have been shown to enhance biodiversity and reduce chemical inputs [173]. Similarly, Napa Valley in the USA has utilized owl boxes and raptor perches to attract birds of prey, which help control rodent populations, while also engaging in habitat restoration projects to support local biodiversity [173]. In South Australia, vineyards in the Barossa Valley and McLaren Vale have incorporated native vegetation and buffer zones to protect against erosion and promote the health of surrounding ecosystems [173].

Despite the benefits of promoting biodiversity, challenges remain in implementing these practices. Some vineyard owners may face higher initial costs and require additional labour, which can deter them from adopting sustainable practices [175]. However, the long-term advantages, such as reduced chemical inputs and enhanced vine health, often outweigh these challenges. Additionally, increasing consumer demand for sustainably produced wines presents an opportunity for vineyards to differentiate their products in a competitive market [176].

Chapter 3
Planning and Designing a Sustainable Vineyard

Choosing the Right Grape Varieties

Wine grape varieties of commercial importance represent only a small subset of the thousands of grapevine varieties cultivated worldwide. Most of the grapes grown for wine production belong to the European vine species *Vitis vinifera*, which has been a cornerstone of winemaking for more than six thousand years. While *Vitis vinifera* dominates the global wine industry, many other grape species are native to America and are also commercially cultivated. Among these, the best-known species is *Vitis labrusca*. Additionally, French-American hybrid varieties, which are crosses between *Vitis vinifera* and native American species, have gained popularity for their resilience and unique qualities.

Vitis vinifera varieties are widely regarded as the finest for producing world-class wines. Each grape variety possesses distinct characteristics, including flavour, colour, berry size, phenolic content, and the balance of sugars and acids. These attributes influence the wine's final profile but are significantly affected by external factors such as terroir—the combination of soil, microclimate, and other environmental conditions in the vineyard. Viticultural management practices, such as pruning, canopy management, and irrigation, also shape the characteristics of the grapes. Furthermore, winemaking techniques, including fermentation methods, aging, and blending, play a significant role in how the grape's qualities manifest in the finished wine.

A cultivated grape variety is formally referred to as a "cultivar." However, the term "variety" is more commonly used in non-technical contexts and is widely recognized in the wine industry. In Europe, the finest wines are typically identified by their regional names rather than the grape variety. For example, Chardonnay and Pinot Noir are the primary grapes associated with Burgundy in France. In contrast, wines from regions like the Americas, Australia, South Africa, and New Zealand are usually labelled by their varietal names, reflecting a different marketing and naming tradition.

When planting new vineyards, growers can further refine their choices by selecting specific clones of a grape variety. Clones are genetic subtypes of a variety, chosen for particular traits such as flavour profile, berry size, cluster shape, vine yields, vigour, and adaptability to environmental challenges like heat, humidity, and drought. These selections allow growers to tailor their vineyards to regional conditions and market demands, enhancing both the quality and resilience of their production.

The interplay between grape variety, terroir, and viticultural practices highlights the complexity and artistry of winemaking. Each decision—from the choice of variety or clone to the management of the vineyard and winemaking techniques—contributes to the unique character of the wine, making it a true reflection of its origins.

Deciding on the grape variety is a complex and critical decision for establishing a vineyard. It requires a thorough evaluation of multiple factors, including the growing method, the suitability of grape varieties to local climatic and soil conditions, and the marketability of the resulting product. The growing method—whether using autogenous plants or grafted cuttings—also plays a significant role, as each method comes with unique benefits and challenges. Additionally, the specific qualities of each grape variety, influenced by factors such as soil pH, electrical conductivity (EC), water and nutrition requirements, and climate conditions, must align with the chosen site. Careful, fact-based selection is essential for long-term success.

Historically, European viticulture predominantly relied on autogenous grapevines—plants that grow on their original root systems. However, this practice faced a catastrophic setback in the mid-19th century with the arrival of *Phylloxera vastatrix*, a devastating soil aphid. Native to America, phylloxera caused widespread destruction in European vineyards, as most *Vitis vinifera* varieties lacked resistance to this pest. American grapevines, having coexisted with phylloxera for centuries, had developed natural immunity.

Today, autogenous grapevines are rarely used in regions where phylloxera is present. However, in areas unaffected by this pest, some smallholder farmers continue to prefer planting autogenous vines. These vines are valued for their purity and traditional characteristics, but their susceptibility to phylloxera makes them a risky choice in most wine-producing regions.

The phylloxera crisis in the 1850s prompted European growers to seek solutions, ultimately turning to American grapevines for their resistance to the pest. The solution involved grafting *Vitis vinifera* varieties onto American rootstocks, combining the desirable fruit characteristics of European grapes with the resilience of American root systems. Grafting involves joining parts from two different plants, where the scion (upper part) grows on the rootstock (lower part), resulting in a single plant that combines the advantages of both components.

Different rootstocks are preferred depending on soil and temperature conditions. Popular rootstocks include *Vitis riparia* (e.g., Riparia Gloire de Montpellier), *Vitis rupestris* (e.g., Rupestris du Lot), and hybrids such as Riparia-Rupestris combinations or Berlandieri-based rootstocks. Each rootstock offers unique morphological and physiological traits, such as drought tolerance, resistance to soil salinity, or adaptability to specific pH levels. The choice of rootstock must align with the vineyard's environmental conditions and desired grape variety.

Regardless of whether autogenous vines or grafted cuttings are chosen, it is critical to purchase plants from reputable sellers to ensure quality and avoid issues like disease or poor graft compatibility.

Grape varieties have been cultivated and refined over thousands of years, resulting in thousands of distinct options. These varieties are generally divided into three main categories based on their primary use: winemaking, currant production, and table grapes.

This category includes mostly European *Vitis vinifera* varieties, renowned for their long-standing association with high-quality wine production. Common red winemaking varieties include Cabernet Franc, Cabernet Sauvignon, Malbec, Merlot, Pinot Noir, Syrah, and Concord. Popular white varieties include Chardonnay, Pinot Blanc, Pinot Gris, Semillon, Gewürztraminer, Catawba, and Delaware. While

European varieties dominate this category, some American varieties also play an important role, particularly in hybrid or regional wines.

Sultanina and Korinth are the most famous varieties in this group, valued for their use in dried fruit production. These varieties are less common in vineyards focused on winemaking but are essential in regions specializing in raisin or currant production.

Table grapes are cultivated for fresh consumption and are distinct from winemaking varieties in terms of flavour, texture, and appearance. Popular varieties include Cardinal, Perlette, Victoria, and Ribier. These grapes are often larger, seedless, and sweeter than those used for wine production.

Selecting a grape variety involves more than just assessing the potential yield or quality; it also requires aligning the variety's characteristics with the vineyard's unique environmental conditions and market demands. For winemaking, understanding the interaction between the chosen variety, terroir, and viticultural practices is essential for producing high-quality, marketable wines. For table grape or currant production, growers must prioritize traits like berry size, sweetness, and resistance to transportation damage.

Ultimately, careful planning, research, and consultation with experts or agricultural extension services can help ensure the chosen varieties and growing methods align with both the environmental conditions of the site and the grower's business goals. This thoughtful approach is key to establishing a successful and sustainable vineyard.

While most commercial grape growers use grafted plants to protect against phylloxera, some regions without this pest still prefer autogenous plants. However, growers must avoid planting a new vineyard immediately after removing an old one from the same location. Recently used vineyard soil is often depleted and potentially infected with pathogens, requiring a fallow period of two to five years before replanting. Consulting a licensed agronomist for advice on the replanting timeline is essential.

Selecting the right grape variety is one of the most critical decisions for vineyard success. Each variety has unique qualities that are only expressed under specific climate, soil, and growing conditions. The compatibility between the rootstock and scion is vital, and the variety must be well-suited to the regional climate. Grapevines generally prefer warm, dry summers and cold but not frosty winters. Ideal soils contain less than 25% clay, a small percentage of gravel, and sufficient organic matter to support healthy growth. Excessive humidity during summer can increase fungal infections, and temperatures below -3°C (27°F) in spring or -15°C (5°F) during dormancy can damage the vine. Soil temperatures should remain above 5°C (41°F) to optimize nutrient uptake, and the optimal pH range for most varieties is between 6.5 and 7.5, though some can tolerate pH levels as low as 4.5 or as high as 8.5.

Once all regulatory and variety selection processes are completed, growers move on to pre-planting preparations. This involves tilling the soil and removing crop remains, though excessive tillage on inclined ground can lead to erosion. On steep slopes, leveling may be necessary to prevent water from pooling in lower areas and creating waterlogged conditions. Drip irrigation systems are typically installed before planting in irrigated vineyards. At transplanting time, growers dig small holes to accommodate the seedlings, followed by fertilization, irrigation, and weed management.

Post-transplanting, growers focus on shaping and training the vines. Various training systems can be employed depending on the grape variety, environmental conditions, soil characteristics, and harvesting techniques. Training involves guiding the vine's growth using support structures and pruning to achieve the desired shape. For winemaking varieties, this process typically takes two to three years, while table grape varieties require one to two years.

Once trellising and vine shaping are completed, growers begin the annual maintenance cycle. This includes pruning, deadheading, defoliation, and grape thinning. Many growers selectively remove developing sprouts throughout the growing season to channel the vine's energy into fewer but higher-quality fruits. This practice, while not universally adopted, is favoured by producers focused on premium quality. During the growing season, vigilant crop monitoring is essential to prevent the spread of diseases and address other potential issues.

Harvesting methods vary depending on the grape's intended use. Wine grapes may be harvested manually using scissors or knives or mechanically with harvesting tractors. Table grapes, however, must be harvested manually to preserve their delicate structure. Traditional vineyards in Europe that produce high-quality, low-yield wines often opt for manual harvesting. The timing of grape harvesting is influenced by the variety, climate, soil characteristics, and growing techniques. It is rarely consistent year to year, even within the same vineyard. Generally, northern hemisphere grapes mature between August and November, while southern hemisphere harvests occur from March to August.

After harvesting, growers meticulously separate healthy grapes from diseased ones, clean them, and decide whether to sell them raw or begin the winemaking process.

Following harvest and leaf drop, the vine enters its dormancy period, conserving energy for the next growing cycle.

Yield expectations vary significantly between table grape and wine varieties. Table grapes typically produce higher yields, ranging from 20 to 50 tons per hectare. In contrast, winemaking varieties often yield less, with premium varieties like Sauvignon or Cabernet limited to around six tons per hectare to maintain quality. While this yield may seem low, premium grapes fetch higher market prices, ensuring profitability for the grower. Medium- to low-quality winemaking varieties can yield 20 to 40 tons per hectare or more but command lower prices in the market. Striking the right balance between quantity and quality is fundamental to long-term success and sustainability in grape growing.

Disease-Resistant Cultivars

Disease-resistant cultivars in viticulture are grapevine varieties specifically bred or selected for their ability to resist or tolerate common diseases that affect grapevines. These cultivars are essential for sustainable viticulture, as they reduce the reliance on chemical treatments such as fungicides, which can be both costly and environmentally harmful [177, 178]. By incorporating genetic resistance to pests and pathogens, disease-resistant cultivars contribute significantly to vineyard health, environmental sustainability, and economic efficiency [179, 180].

The development of disease-resistant cultivars involves breeding programs that combine desirable traits from traditional grape varieties (Vitis vinifera) with resistance traits from other grape species, such as Vitis labrusca, Vitis riparia, Vitis aestivalis, and Vitis rotundifolia. This process can involve traditional cross-breeding or advanced genetic techniques, allowing these cultivars to inherit resistance while maintaining the quality traits necessary for winemaking or table grape production [181, 182]. For instance, the identification of resistance loci in wild grape species has been key for breeding programs aimed at enhancing disease resistance in cultivated varieties [183, 184].

Key diseases targeted by resistance breeding include powdery mildew (Erysiphe necator), downy mildew (Plasmopara viticola), Botrytis bunch rot (Botrytis cinerea), black rot (Guignardia bidwellii), and phylloxera (Daktulosphaira vitifoliae). Resistant cultivars can significantly reduce the need for chemical fungicides, particularly in humid climates where these diseases thrive [185, 186]. For example, cultivars like Regent and Seyval Blanc have shown resistance to multiple diseases, thereby improving vineyard health and fruit quality [187, 188]. The ability of these cultivars to thrive in challenging conditions not only enhances yield but also supports organic and biodynamic practices by minimizing chemical inputs [189, 190].

Despite the advantages, challenges remain in the adoption of disease-resistant cultivars. Some hybrids may not produce wines with the same complexity as traditional Vitis vinifera varieties, which can affect market acceptance [191, 192]. Additionally, the success of these cultivars often depends on their compatibility with local climate and soil conditions, necessitating thorough evaluation and trials before widespread planting [193, 194]. Moreover, the durability of resistance is a concern, as pathogens can evolve to overcome resistance traits, highlighting the need for ongoing research and development in this area [195, 196].

Matching Grapes to Climate and Soil

Matching grape varieties to the climate and soil of a vineyard is a foundational step in successful viticulture. The interplay between the grapevine, its environment, and vineyard management practices significantly influences the vine's health, the quality of the grapes, and the character of the resulting wine. By understanding the unique climate and soil characteristics of a site, growers can select grape varieties and rootstocks best suited to thrive under specific conditions, ensuring optimal yields and sustainability.

Climate is the most critical factor in determining which grape varieties will thrive in a particular region. Grapevines are sensitive to temperature, rainfall, sunlight, and humidity, all of which affect their growth, ripening, and vulnerability to diseases.

Temperature plays a pivotal role, with each grape variety requiring specific growing season averages. Cool-climate varieties such as Pinot Noir and Riesling excel in regions with mean temperatures of 13–15°C (55–59°F), as these conditions allow them to develop their signature high acidity and nuanced flavours. In contrast, warm-climate varieties like Shiraz and Cabernet Sauvignon perform better in regions with averages of 18–20°C (64–68°F), where they can achieve full ripeness and bold flavour profiles. However, extreme heat can result in over-ripening and reduced acidity, while prolonged low temperatures can stunt ripening and sugar accumulation.

Rainfall is another crucial factor. Grapevines need adequate rainfall during dormancy and the growing season, but excessive rain during ripening can dilute flavours and increase fungal disease risks. Drought-tolerant varieties like Grenache and

Tempranillo are ideal for drier climates, while mildew-resistant hybrids such as Seyval Blanc and Chambourcin are better suited to humid areas.

Sunlight is equally vital, as it drives photosynthesis and influences the accumulation of sugars and phenolic compounds. South-facing slopes in the Northern Hemisphere (or north-facing in the Southern Hemisphere) maximize sunlight exposure, making these orientations ideal for varieties requiring high light intensity. Wind and humidity also factor into variety selection, as high humidity increases fungal risks and wind can either enhance airflow for disease control or damage vines if too strong.

Soil quality and composition have a profound effect on vine vigour, root development, and fruit quality. Matching soil properties to the needs of specific grape varieties ensures a robust root system and optimal growth conditions.

Drainage is a critical soil characteristic. Well-drained soils, such as sandy or gravelly soils, prevent waterlogging and root rot, making them ideal for deep-rooting varieties like Cabernet Sauvignon. In contrast, clay soils that retain more moisture may be suitable for drought-prone regions. Fertility is another consideration; moderately fertile soils encourage balanced vine growth, while excessively rich soils can lead to excessive vigour and diminished fruit quality. Low-fertility soils, such as limestone or slate, help concentrate grape flavours and are prized for producing high-quality wines from varieties like Syrah and Chardonnay.

The soil's pH significantly impacts nutrient availability. Most grapevines thrive in soils with a pH between 6.0 and 7.5, though some varieties, such as Concord, can tolerate acidic soils, and certain Riesling clones can adapt to slightly alkaline conditions. Texture also matters, as sandy soils promote quick drainage, clay soils retain moisture, and loamy soils balance both properties, making them ideal for most grape varieties. Finally, mineral content, particularly calcium, magnesium, and potassium, affects vine health and grape quality. Limestone-rich soils, for instance, are associated with crisp, high-acid wines like those from Chardonnay or Sauvignon Blanc.

The choice of rootstock is integral to matching grapes to specific environments. Rootstocks provide tolerance to various soil conditions, pests, and diseases. For regions with phylloxera, rootstocks derived from American species such as *Vitis riparia* or *Vitis rupestris* are indispensable. Rootstocks like Riparia Gloire de Montpellier are suited to high-fertility soils, while hybrids like Berlandieri x Rupestris thrive in alkaline or drought-prone conditions. Rootstocks with deep roots are ideal for arid regions, while shallow-rooting options perform well in consistently moist soils.

In cool-climate regions like Germany's Mosel Valley or New Zealand's Marlborough, the combination of cool temperatures and slate or loamy soils supports high-acid varieties such as Riesling or Sauvignon Blanc. Warm-climate regions like Spain's Rioja or Australia's Barossa Valley, with their warm temperatures and well-drained soils, are ideal for robust varieties like Tempranillo, Grenache, and Shiraz. Arid areas such as California's Paso Robles or southern Spain favour drought-resistant varieties like Grenache and Mourvèdre, while high-humidity regions such as the eastern United States are well-suited for hybrids like Seyval Blanc or Chambourcin, which resist fungal diseases.

Before planting, conducting thorough soil and climate assessments is essential. Soil testing provides critical data on pH, fertility, and texture, while climate data offers insights into temperature ranges, rainfall patterns, and sunlight exposure. This information helps growers make informed decisions about the most suitable grape varieties and rootstocks for their site. Consulting with viticultural experts or agricultural extension services further ensures that decisions are tailored to the unique conditions of the vineyard.

Vineyard Layout

Key aspects of vineyard layout include row orientation, spacing between rows and vines, slope and drainage management, trellis and training systems, irrigation infrastructure, access pathways, biodiversity areas, windbreaks, zoning for microclimates, and storage infrastructure.

Figure 12: Photo illustrating the small spacing between vine aisles in Burgundy. The small space means that most of the harvesting and pruning is done by hand and not mechanical. Philip Larson, CC BY-SA 2.0, via Wikimedia Commons.

The orientation of vineyard rows is critical for maximizing sunlight exposure and minimizing disease risk. Typically, rows are aligned north-to-south to ensure uniform sunlight distribution, which is essential for balanced ripening of grapes. In cooler climates, rows may be angled to capture more sunlight, while in hotter regions, east-to-west orientations can provide shade during peak sun hours, thus preventing heat stress on the vines [155, 197]. This strategic orientation not only enhances grape quality but also contributes to sustainable vineyard practices by reducing the need for chemical treatments against diseases that thrive in shaded, damp conditions [198].

Row and vine spacing is influenced by grape variety, rootstock, soil fertility, and the chosen training system. For small-scale vineyards, a balance must be struck between maximizing land use and ensuring adequate airflow and light penetration. Standard row spacing ranges from 2 to 3 meters (6.5 to 10 feet) to facilitate equipment access and airflow, while vine spacing typically varies from 1 to 1.5 meters (3 to 5 feet) depending on the vigour of the grape variety [199, 200]. Proper spacing is imperative for preventing overcrowding, which can lead to increased fungal diseases and competition for water and nutrients [201].

In sloped vineyards, layout considerations must include effective drainage and erosion control. Aligning rows along the slope's contour can significantly reduce water runoff and soil erosion, while terracing may be necessary on steeper slopes to create level planting surfaces [197]. In flat vineyards, implementing drainage systems, such as ditches or tiles, is essential to prevent waterlogging, particularly in heavy or clay-rich soils [198, 200]. These practices not only enhance vine health but also contribute to the sustainability of vineyard operations by maintaining soil integrity and reducing erosion.

Figure 13: Small scale sloped vineyard, Durbach Black Forest. pixabay.com, CC0, via Picryl.

The choice of trellis and training systems plays a vital role in vineyard layout. Common systems in small-scale vineyards include Vertical Shoot Positioning (VSP), Guyot (single or double), and Goblet (head training) systems. Each system has its advantages, such as promoting good airflow and sunlight penetration, which are important for vine health and grape quality [199]. The selection of an appropriate system depends on various factors, including grape variety, climate, and labour availability, thereby necessitating careful planning and consideration [200].

In regions with inconsistent rainfall, integrating a drip irrigation system into the vineyard layout is often essential. Drip irrigation conserves water by delivering it directly to the roots, thus minimizing evaporation losses [201]. Additionally, small-scale vineyards may benefit from rainwater harvesting systems to supplement irrigation needs, enhancing water use efficiency and sustainability [198, 200]. The design of irrigation infrastructure must be carefully planned to ensure that all vines receive adequate water, which is particularly important in arid regions where water scarcity is a significant concern [202].

Access pathways are necessary for vineyard management and harvesting. In small-scale vineyards, pathways should be designed to facilitate easy movement of workers and equipment while maintaining a balance between land use and operational efficiency [155]. Strategic placement of access points can enhance the efficiency of maintenance tasks, such as pruning and spraying, thereby improving overall vineyard productivity [198].

Incorporating buffer zones or biodiversity areas within vineyard layouts promotes sustainability and reduces environmental impact. These areas can include native vegetation, hedgerows, or cover crops, which enhance soil health, attract beneficial insects, and decrease the reliance on chemical inputs [200]. Such practices not only contribute to the ecological balance but also support the vineyard's long-term viability by fostering a healthier growing environment for the vines.

Windbreaks, such as rows of trees or shrubs, are essential in regions prone to strong winds. These barriers protect vines from wind damage and reduce evaporation rates, thereby enhancing vine health and grape quality [155, 200]. The careful positioning of windbreaks is important to avoid shading the vines while still providing effective protection [198].

Zoning within vineyards allows for the management of microclimatic variations, which can significantly influence grape quality. For instance, cooler areas may be planted with varieties that thrive in less heat, while warmer zones can host heat-tolerant varieties [155, 200]. This tailored approach to vineyard management optimizes grape quality and yield by aligning grape varieties with their preferred growing conditions [198].

Figure 14: Grapevines in a greenhouse. Greece-China News. CC0, via Pexels.

Even small-scale vineyards require designated areas for storage and essential infrastructure, including tools, fertilizers, and irrigation equipment. In vineyards focused on winemaking, space for fermentation tanks and wine storage is also necessary [155, 200]. Proper planning of these facilities ensures that operational efficiency is maintained while supporting the vineyard's overall productivity.

Growing Grape Vines in a Greenhouse

Cultivating grapevines in a greenhouse offers the potential for exceptional quality and flavour, making it a preferred method for growers seeking consistent results, especially in cooler climates. A greenhouse provides a controlled environment that protects the vines from adverse weather and pests, ensuring better growth conditions and higher fruit quality. Whether using a lean-to, conservatory, sun lounge, or traditional greenhouse, the principles of indoor grapevine cultivation remain consistent. However, cultivating vines inside a house or room, even with excellent light levels, is not feasible due to the excessively dry indoor atmosphere.

For those living in northern or cooler climates, greenhouse grapevines often outperform their outdoor counterparts in terms of quality and yield. While some grape varieties are suited for outdoor cultivation in such regions, the controlled environment of a greenhouse allows for better temperature regulation, reduced frost risk, and enhanced sunlight exposure, all of which are critical for ripening fruit and developing sugars.

There are two primary methods for planting grapevines in a greenhouse:

1. Planting Roots Outside the Greenhouse

In this method, the vine's root system is planted outside the greenhouse, against its wall, with the vine trained through a small hole into the interior. This approach leverages natural rainfall to water the roots, reducing the need for manual irrigation. However, it has some limitations:

- The soil outside tends to be colder, especially in early spring, which can slow growth and delay fruiting.

- Natural watering may not be sufficient during dry spells, requiring supplemental irrigation.

2. Planting Roots Inside the Greenhouse

Alternatively, the vine can be planted directly into a border inside the greenhouse or in a container. This method offers several advantages:

- Warmer soil temperatures inside the greenhouse encourage faster and more vigorous growth.

- Growth tends to be more profuse, potentially leading to earlier and larger harvests.

However, planting inside requires meticulous attention to watering, as there is no natural rainfall to sustain the vine. Regular irrigation is essential to prevent the roots from drying out, especially during the growing season.

Whether planting indoors or outdoors, providing the right soil conditions is crucial for vine health. In most cases, growers will need to import planting soil for greenhouse cultivation. The ideal soil mix is porous, well-draining, and nutrient-rich. A recommended formula includes:

- 60% loam for structure and nutrients.

- 20% peat to retain moisture and provide organic matter.

- 20% coarse grit to improve drainage and aeration.

For added fertility, a dusting of bonemeal can be incorporated into the mix. Alternatively, ready-made compost, such as John Innes No. 2, can be used with additional peat and grit to enhance its properties. Regardless of the planting medium, ensuring excellent drainage is critical to avoid waterlogging and root rot.

The positioning of the grapevine within the greenhouse is vital for maximizing sunlight exposure. Grapevines require abundant sunlight to ripen their fruit properly, achieve vibrant colour, and develop higher sugar levels. Even in small greenhouses, vines can be trained to fit available space without compromising productivity. Historically, grapevines were allowed to grow freely and form large, sprawling structures for aesthetic and practical reasons. However, modern cultivation methods emphasize proper pruning and training to maintain manageable vine sizes and focus the plant's energy on fruit production rather than excessive vegetative growth.

Pruning is a critical practice in greenhouse grapevine cultivation. It keeps the vine within manageable proportions, encourages fruiting, and prevents the plant from becoming overly vegetative. Regular pruning ensures that sunlight and airflow reach all parts of the vine, reducing the risk of fungal diseases and promoting even ripening. This approach mirrors the methods used in commercial vineyards, where efficient use of space and resources is paramount.

Greenhouse cultivation allows growers to create a microclimate that optimizes vine health and grape quality. The controlled environment mitigates many challenges faced by outdoor vineyards, such as frost, pests, and inconsistent weather patterns. Additionally, even small structures can accommodate grapevines, making this method accessible to hobbyists and small-scale producers alike. By employing proper pruning techniques, monitoring irrigation, and ensuring adequate sunlight, growers can achieve impressive results, regardless of the greenhouse size.

Cultivating grapevines in a greenhouse offers growers a controlled environment to optimize vine growth, improve fruit quality, and overcome challenges posed by climate. This method is especially advantageous in regions with harsh winters, unpredictable rainfall, or shorter growing seasons. Whether in the Northern or Southern Hemisphere, the principles of greenhouse viticulture are broadly similar, but specific considerations, such as sunlight angles, seasonal cycles, and training methods, must be adapted to the hemisphere.

Greenhouses create a protective microclimate, shielding vines from external stresses like frost, excessive rain, and pests. This controlled environment is particularly beneficial in regions with extreme temperature variations or short growing periods. In cooler climates, such as northern Europe or southern New Zealand, greenhouses extend the growing season, enabling varieties that struggle outdoors to flourish. Conversely, in warmer climates like southern Australia or California, greenhouses moderate excessive heat and protect vines from unexpected weather events.

Greenhouses vary in form, from standalone structures to lean-tos and conservatories, but their purpose remains the same: to provide a space with adequate light, warmth, and humidity to mimic optimal outdoor conditions.

Planting Roots Outside the Greenhouse: One common method involves planting the vine's roots outside the greenhouse wall and training the top growth through a small hole into the greenhouse interior. This technique leverages natural rainfall to water the roots, reducing irrigation needs. The roots benefit from the consistent moisture provided by natural precipitation, which simplifies maintenance. However, soil temperatures outside the greenhouse tend to be colder, especially in early spring, which can slow root activity and vine growth. This method is most suitable for regions with mild climates and reliable rainfall.

Planting Roots Inside the Greenhouse: Alternatively, planting the roots directly inside the greenhouse, either in a border or a container, offers distinct advantages. Warmer soil temperatures inside the greenhouse promote faster root development and earlier vine growth. Controlled soil conditions also allow for more vigorous growth and potentially higher yields. However, this method requires meticulous attention to watering since the roots do not benefit from natural rainfall. Consistent irrigation is essential, especially during peak growth periods or dry spells.

Regardless of the planting method, preparing the soil is critical to vine health and productivity. In greenhouse cultivation, natural topsoil is often limited, requiring growers to import or amend soil to meet the specific needs of grapevines.

A porous soil mix, consisting of 60% loam, 20% peat, and 20% coarse grit, ensures proper drainage and aeration. Adding organic fertilizers or a dusting of bonemeal supports initial vine growth and long-term fertility. Alternatively, commercial composts like John Innes No. 2, enhanced with peat and grit, provide a ready-made option. Drainage is a top priority, as waterlogging can lead to root rot. Monitoring soil pH is also essential, with most grapevines thriving in a pH range of 6.0–7.5, although some varieties tolerate more extreme levels.

Greenhouse placement and vine orientation are decisive for maximizing sunlight exposure, which drives photosynthesis and ensures proper fruit ripening. In the Northern Hemisphere, greenhouse vines benefit most from a south-facing orientation, while in the Southern Hemisphere, a north-facing orientation achieves the same effect. This positioning ensures even light distribution throughout the day.

During summer, the sun's high trajectory provides ample light, but in winter, the lower sun angle requires careful placement of the vine to avoid shading from nearby structures or vegetation. In regions with limited winter sunlight, reflective materials or supplemental lighting may be necessary to maintain optimal conditions for vine growth.

Greenhouse grapevines require proper training and pruning to maximize space efficiency and encourage fruit production. These practices are especially important in the confined space of a greenhouse.

- **Vertical Shoot Positioning (VSP)**: This system trains shoots upward along wires, improving sunlight exposure and airflow. It is particularly effective in small greenhouses, where maximizing vertical space is essential.

- **Guyot Training**: Single or double Guyot systems involve leaving one or two fruiting canes while pruning excess growth. This method controls vine size and focuses energy on producing high-quality fruit.

- **Traditional Rambling**: Historically, vines were allowed to grow freely in large greenhouse structures, creating visually stunning but less efficient systems. Modern growers typically avoid this method due to its tendency to encourage excessive vegetative growth at the expense of fruit quality.

Regional Considerations:

Northern Hemisphere: In cooler regions like northern Europe or the northeastern United States, greenhouses are invaluable for extending the growing season. Early-ripening varieties such as Pinot Noir and Riesling thrive in controlled environments with moderate temperatures. Insulating the greenhouse and incorporating heating systems during late winter and early spring protect dormant vines from frost damage.

Southern Hemisphere: In regions like southern Australia or South Africa, where summers can bring extreme heat, greenhouses moderate temperature fluctuations and protect against heat stress. Heat-tolerant varieties such as Shiraz and Grenache perform well in these environments. Cooling techniques like shading nets or evaporative cooling systems may be employed to maintain ideal temperatures inside the greenhouse.

Maintaining appropriate humidity levels is essential for greenhouse-grown grapevines. The enclosed environment can become too dry or too humid, both of which pose challenges to vine health. Drip irrigation is the most efficient watering method, delivering water directly to the roots and minimizing evaporation loss. Ventilation systems help control excess humidity, reducing the risk of fungal diseases like powdery mildew. Regular monitoring of soil moisture and air humidity ensures optimal conditions for vine growth.

Greenhouse cultivation of grapevines allows growers to manage environmental factors with precision, leading to high-quality fruit and successful harvests even in challenging climates. By choosing the appropriate planting method, preparing soil carefully, and optimizing sunlight exposure, growers can create ideal conditions for grapevines. Training and pruning ensure efficient use of greenhouse space, while adjustments for hemispheric differences account for seasonal and solar variations. With diligent management, greenhouse viticulture provides a versatile and effective solution for growers aiming to produce exceptional grapes tailored to their specific environment.

Companion Planting and Cover Crops

Companion planting in vineyards is an increasingly recognized practice that enhances the health and productivity of grapevines. This method involves strategically selecting and cultivating companion plants alongside grapevines to create a synergistic environment that promotes vine growth, improves soil health, and provides natural pest control. The integration of companion plants can lead to healthier vineyards, reduced reliance on chemical inputs, and a more sustainable approach to viticulture.

One of the primary benefits of companion planting is the improvement of soil health and structure. Leguminous plants, such as clover and vetch, are particularly effective as they fix atmospheric nitrogen, enriching the soil and reducing the need for synthetic fertilizers [203]. Additionally, deep-rooted species, including native forbs and wildflowers, can alleviate soil compaction, enhancing aeration and water infiltration, which is required for grapevine root development. This practice not only supports the immediate nutrient needs of grapevines but also contributes to long-term soil fertility.

Natural pest control is another significant advantage of companion planting. Aromatic plants like lavender and rosemary can repel harmful pests due to their strong scents while also attracting beneficial insects that serve as natural predators. For instance, lavender has been shown to deter aphids and other vineyard pests, thereby fostering a balanced ecosystem that minimizes the need for chemical pesticides. The presence of diverse plant species can enhance the overall resilience of the vineyard against pest outbreaks.

Moreover, companion plants can enhance pollination, which is vital for the fruit set of self-pollinating grape varieties. Flowering plants attract pollinators such as bees, thereby improving the consistency of fruit set and potentially increasing yields. This is particularly important in vineyards where pollination can directly influence the quality and quantity of grape production.

Companion plants also play a significant role in weed suppression. Ground cover species like clover and certain grasses can act as natural mulch, effectively suppressing weed growth and conserving soil moisture. This reduces competition for resources between grapevines and weeds, thereby enhancing grapevine health. Furthermore, the biodiversity introduced through companion planting creates a more resilient ecosystem capable of withstanding environmental stressors such as drought and disease.

The incorporation of diverse plant species not only supports pest management but also fosters a rich microbial community in the soil. This increased biodiversity can lead to improved nutrient cycling and enhanced soil health, which are essential for sustainable vineyard management. Studies have indicated that greater plant diversity correlates with reduced pest damage and improved ecosystem services, highlighting the importance of biodiversity in agricultural systems.

Several specific companion plants have been identified as particularly beneficial for grapevines. Herbs such as rosemary and thyme not only stabilize soil moisture but also repel pests and attract pollinators. Flowers like chamomile and marigolds can combat mildew and deter harmful insects, respectively, while vegetables such as garlic and onions provide additional pest-repelling properties. Cover crops like alfalfa and buckwheat contribute to soil health and attract beneficial insects, further enhancing the vineyard ecosystem.

Effective companion planting requires careful planning and monitoring. Intercropping strategies, where companion plants are planted between vineyard rows or within rows, can maximize benefits without compromising grapevine health. Proper spacing is essential to ensure that both grapevines and companion plants have adequate access to resources, preventing competition that could hinder growth. Common mistakes to avoid include overcrowding companion plants and selecting incompatible species that may inhibit grapevine growth or attract pests.

Infrastructure for Sustainability

Irrigation

Irrigation in viniculture involves the supplemental application of water to grapevines during their growing season. It is both a controversial and essential practice, depending on the region, climate, and viticultural philosophy. Water availability significantly impacts the physiological processes of the grapevine, influencing photosynthesis, vine growth, and the development of grape berries. While sufficient water is necessary to avoid vine stress, strategic water management can also enhance grape quality by concentrating flavours and sugars.

A typical grapevine requires 25-35 inches (635-890 millimetres) of water annually, primarily during the spring and summer months. This water supports photosynthesis, the process by which the vine produces the energy necessary for growth and fruit development. Inadequate water during these critical periods results in water stress, which alters vine physiology. While excessive water stress can impede photosynthesis and nutrient storage, moderate stress can produce desirable effects for wine grape growers, such as smaller berry size, higher sugar concentration, and more intense flavours. These characteristics are particularly valued for crafting high-quality wines.

In many traditional Old World wine regions, natural rainfall is considered the only acceptable source of water for grapevines. This philosophy ties closely to the concept of terroir—the unique combination of soil, climate, and geography that influences wine characteristics. Critics of irrigation argue that it undermines terroir by artificially altering the natural growing conditions, potentially leading to excessive yields and diluted wine quality. As a result, irrigation has historically been banned under European Union wine laws.

However, attitudes toward irrigation are gradually shifting in some European countries. For instance, Spain has relaxed regulations, allowing limited irrigation to support viticulture in its arid regions. Similarly, France's wine regulatory body, the Institut National des Appellations d'Origine (INAO), has begun reviewing irrigation policies to address the realities of climate change and modern viticultural practices.

In contrast to Old World traditions, irrigation is widely accepted and often essential in many New World wine regions, such as Australia, California, and parts of South America. These regions frequently experience dry climates with minimal rainfall, making irrigation a critical tool for sustaining viticulture. Research and innovation in these areas, as well as in arid Old World regions like Israel, have demonstrated that carefully managed irrigation can enhance wine quality.

The principle of controlled water stress plays a key role in these practices. During the early growing season, when vines are budding and flowering, sufficient water is provided to support initial growth. As the ripening period approaches, irrigation is scaled back to create moderate water stress. This stress signals the vine to allocate more resources to developing grape clusters rather than excessive foliage, concentrating sugars, flavours, and phenolics in the berries. This approach ensures a balance between vine health and optimal grape quality.

While irrigation provides significant benefits, it also comes with risks. Excessive water stress can lead to a shutdown of essential vine processes, such as photosynthesis, nutrient absorption, and growth. On the other hand, over-irrigation can dilute flavours, increase yields excessively, and promote diseases associated with high humidity. The availability of irrigation systems offers growers the flexibility to respond to drought conditions and maintain a balance between water stress and vine development.

Irrigation systems are a critical component of vineyard management, representing a significant investment that requires careful planning and design. Several factors influence the selection of an appropriate irrigation system, including soil type, depth, and the effective rooting zone of the vines, as well as vine density, water quality, land formation, and economic considerations such as capital and operational costs. An efficient irrigation system supports healthy vine growth while minimizing water wastage and environmental impacts like soil erosion. Water losses from evaporation, wind drift, runoff, or deep percolation below the root zone can significantly reduce irrigation efficiency. Understanding the various irrigation systems available for vineyards allows growers to select the best approach for their specific circumstances.

The choice of irrigation system must align with the vineyard's unique requirements. The soil type and depth determine how water is retained and distributed, while the effective rooting zone of the vines dictates the volume and timing of water applications. Vine density affects the water demand, and the vineyard's topography influences water distribution methods. Additionally, water quality—particularly salinity—can impact both the vines and soil health, requiring systems that mitigate potential negative effects. Economic factors, such as the upfront cost of installation and ongoing operational expenses, also play a role in system selection. The overarching goal is to optimize water use, promote vine health, and sustain long-term vineyard productivity.

Different irrigation systems cater to diverse vineyard needs, each with distinct advantages and limitations. The primary options include overhead sprinkler irrigation, drip irrigation, furrow irrigation, and micro-sprinkler irrigation. Each system is tailored to specific vineyard conditions, and no single solution fits all scenarios.

Overhead sprinkler irrigation systems distribute water across the vineyard using high-pressure sprinklers mounted on permanent standpipes that rise above the vines. This method effectively simulates natural rainfall, evenly distributing water over large areas. Overhead sprinklers are particularly useful in frost-prone regions, as they can double as frost protection systems by forming a protective layer of ice on the vines during freezing conditions.

However, these systems are water-intensive and can result in significant losses due to evaporation and wind drift. Soil erosion may also occur if water application rates exceed the soil's infiltration capacity. Despite these drawbacks, overhead sprinkler irrigation remains a viable choice for large-scale vineyards where uniform water distribution is critical.

Drip irrigation, also known as trickle irrigation, is one of the most water-efficient methods for vineyard irrigation. This system delivers water directly to the root zone at very low rates through a network of small-diameter plastic pipes fitted with emitters or drippers. Emitters come in two configurations: compensating emitters, which maintain a consistent flow rate despite changes in pressure, and non-compensating emitters, which vary flow rates depending on pressure.

Drip irrigation minimizes water loss through evaporation and runoff, ensuring precise water delivery to the vines. It also reduces weed growth and the risk of fungal diseases by keeping the foliage dry. However, the system requires careful monitoring and maintenance to prevent clogging in the emitters, especially in regions with hard water or high sediment levels. Drip irrigation is highly suited for vineyards in arid or semi-arid regions where water conservation is paramount.

Furrow irrigation involves channelling water from a main ditch into furrows that run along either side of the vine rows or through a central channel. This gravity-fed method is simple and cost-effective, making it a popular choice for vineyards with access to abundant surface water and suitable topography.

Despite its simplicity, furrow irrigation is less efficient than other methods. Water losses through evaporation, deep percolation, and runoff are common, and achieving uniform water distribution can be challenging. Soil erosion and salinity issues may also arise if the system is not carefully managed. Furrow irrigation is best suited for flat or gently sloping vineyards with heavy soils that retain water well.

Micro-sprinkler irrigation systems are similar to drip irrigation but use small sprinklers to spray water over a localized area rather than delivering it directly to the soil. These systems are made from plastic and come in various flow rates and spray patterns, allowing growers to customize water distribution to the specific needs of their vineyard.

Micro-sprinklers provide uniform water coverage, making them ideal for young vines with shallow root systems or vineyards where maintaining a moist soil surface is important. They also help reduce water loss due to evaporation compared to overhead sprinklers, although some loss may still occur. Micro-sprinklers are particularly effective in sandy soils that require frequent but moderate water applications.

Drip Irrigation

Drip irrigation has become the most widely used method of watering in vineyards due to its precision, efficiency, and adaptability to various soil types, terrains, and environmental conditions. Compared to other irrigation systems like furrow or overhead sprinklers, drip irrigation minimizes water loss, reduces soil erosion, and provides consistent water delivery to grapevines. This approach is particularly valuable in modern viticulture, where water conservation and optimized grape quality are paramount.

Figure 15: Vineyard Drip Irrigation. Jeff Vanuga, USDA Natural Resources Conservation Service, U.S. Department of Agriculture, Public Domain, via Picryl.

Drip irrigation addresses many challenges faced by vineyard managers, including variability in soil composition, topography, and water availability. It allows precise control over water distribution and irrigation timing, ensuring that water is applied directly to the root zone of each vine. This minimizes losses due to evaporation, runoff, or deep percolation, making it more efficient than alternative irrigation methods. Drip systems are also highly compatible with automation, reducing labour costs and enabling better scheduling based on real-time field conditions.

The flexibility of drip irrigation makes it ideal for vineyards facing water supply shortages, high irrigation costs, or elevation differences within the field. By tailoring the system design to vineyard-specific needs, managers can address soil permeability and variability while maintaining optimal vine hydration.

A typical vineyard drip irrigation system includes pumps, filters, chemical injectors, main and submain lines, laterals, and emitters. Water sources are often underground wells, which may deliver water directly to the field or store it in tanks or reservoirs. From these storage points, water can be distributed through gravitational flow or pressurized by booster pumps.

At the pump station, filtration devices and chemical injection systems are critical for maintaining water quality and preventing emitter clogging. Flow meters and pressure gauges are installed to monitor system performance and ensure consistent water delivery. Drip hoses, equipped with emission devices (emitters), are laid along the vineyard rows. These emitters deliver water at precise flow rates—commonly 0.5 to 1.0 gallons per hour—directly to the vines' root zones. Pressure-compensating emitters ensure uniform water application, even in fields with elevation changes.

Drip systems can irrigate the entire vineyard simultaneously or in designated sections, known as irrigation sets. This sectional approach provides flexibility for larger vineyards and allows targeted water delivery based on vine needs.

Two key metrics—Irrigation Efficiency (IE) and Distribution Uniformity (DU)—measure the effectiveness of a drip irrigation system. IE evaluates how much of the applied water is used beneficially within the vineyard, with an ideal value of 100%. In contrast, DU assesses how evenly water is distributed across the field. A high DU ensures that all vines receive adequate water, avoiding under- or over-irrigation in specific areas.

DU is calculated using field observations of water application rates, emitter performance, and system pressure. Factors affecting DU include pressure variations, emitter clogging, and uneven spacing or drainage. For optimal performance, vineyard drip systems are designed with DU values of 0.90 or higher. Regular maintenance and monitoring are essential to sustain these levels over time.

Sample Calculations:

Distribution Uniformity (DU):

Distribution Uniformity measures how evenly irrigation water is distributed across a vineyard or field. It is calculated using the ratio of the average depth of water applied to the lowest quarter of the field to the average depth applied across the entire field.

Data Used:

Water depth measurements from different locations in the vineyard (in millimetres): 12,10,15,9,11,10,14,13,10,9,8,11

Steps:

1. Identify the Lowest Quarter of Values:

- Sort the water depths in ascending order: 8,9,9,10,10,10,11,11,12,13,14,15.

- Select the lowest quarter of the sorted data (the smallest 25% of values): 8,9,9.

2. Calculate the Average of the Lowest Quarter:

- Add the lowest quarter values: 8+9+9=26.

- Divide by the number of values: 26/3=8.67.

3. Calculate the Average of All Measurements:

- Add all water depths: 12+10+15+9+11+10+14+13+10+9+8+11=132 mm.

- Divide by the total number of measurements: 132/12=11.0 mm.

4. Calculate DU:

- Use the formula:

$$DU = \left(\frac{\text{Average Low Quarter Depth}}{\text{Average Depth of All Measurements}} \right) \times 100$$

- Substituting the values:

$$DU = \left(\frac{8.67}{11.0}\right) \times 100 = 78.79\%.$$

Irrigation Efficiency (IE):

Irrigation Efficiency evaluates how effectively the applied water is used for beneficial purposes, such as crop growth. It is expressed as a percentage of the water beneficially used to the total water applied.

- **Data Used:**

    o Total water applied: 1000 litres.

    o Water beneficially used for crop growth: 800 litres.

- **Steps:**

1. **Calculate IE:**

- Use the formula:

$$IE = \left(\frac{\text{Water Beneficially Used}}{\text{Total Water Applied}}\right) \times 100$$

- Substituting the values:

Interpretation of Results:

- **DU of 78.79%:** This indicates a moderate level of uniformity in water distribution. Ideally, DU should be closer to or above 90% in a well-designed and maintained system.

- **IE of 80.0%:** This shows that 80% of the applied water is being beneficially used for crop growth, with the remaining 20% potentially lost to evaporation, runoff, or deep percolation.

By improving DU (e.g., through better emitter performance or pressure management), overall irrigation efficiency can be further optimized.

Drip irrigation systems require routine maintenance to prevent performance degradation. Over time, emitters can become clogged due to insufficient filtration, debris accumulation, or the use of organic fertilizers. Pressure regulators and valves may also wear out, leading to uneven water distribution.

Regular inspections should include:

1. **Filter Maintenance**: Filters should be cleaned or replaced as needed to prevent clogging. Automated backflush systems can help maintain filter performance.

2. **Pressure Monitoring**: Pressure gauges installed at key points in the system should be checked frequently to identify pressure losses caused by blockages or equipment malfunctions.

3. **Emitter Maintenance**: Emitters should be inspected for clogging or wear. Flushing downstream hoses can remove debris and ensure consistent water flow.

4. **Valve and Regulator Servicing**: Pressure-regulating valves (PRVs) should be calibrated and adjusted periodically to maintain uniform water distribution. Damaged or non-adjustable valves should be replaced with high-quality alternatives.

A thoughtfully designed drip irrigation system is essential for ensuring the long-term sustainability and efficiency of vineyard operations. Key factors such as soil type, topography, vine spacing, and water quality must be carefully considered during the design phase. These elements play a role in determining how well the system functions and its ability to support consistent vine growth and grape production.

Drip irrigation systems are typically designed to perform effectively for up to 20 years. However, achieving this longevity requires regular maintenance, vigilant monitoring, and adherence to best practices. High-quality components are the foundation of a durable system, as they resist degradation over time and minimize the risk of malfunctions. Selecting durable materials for emitters, hoses, and pumps is an investment that reduces the likelihood of costly repairs or replacements.

Monitoring water quality is another critical aspect of maintaining system efficiency. Vineyard managers should implement filtration and chemical treatments to prevent emitter clogging and ensure uniform water distribution. Proper filtration not only extends the life of the system but also safeguards the health of the grapevines by preventing contaminants from reaching the root zone.

The design and maintenance of the pump station are equally vital. Pump stations must be optimized to avoid pressure losses and ensure the consistent delivery of water and nutrients. An efficiently designed pump station also facilitates accurate chemical injection, which is necessary for maintaining soil health and vine productivity.

To ensure long-term performance, vineyard managers should regularly evaluate the system's functionality using metrics such as Distribution Uniformity (DU) and Irrigation Efficiency (IE). These measurements provide insights into how evenly and effectively water is being distributed across the vineyard. Promptly addressing any issues identified during evaluations—such as pressure imbalances or emitter malfunctions—prevents small problems from escalating into significant inefficiencies.

By focusing on these design and maintenance principles, vineyard managers can create a drip irrigation system that not only meets immediate irrigation needs but also supports the vineyard's health and productivity for decades.

While drip irrigation offers significant advantages in water efficiency and precision, it is not without its challenges. Common issues such as elevation changes, soil variability, and emitter clogging can compromise the system's efficiency and effectiveness, potentially impacting vine health and grape yield. However, these challenges can be addressed with targeted strategies and proactive management.

Elevation changes within a vineyard can lead to uneven water distribution, with lower areas receiving more water due to gravity and higher areas potentially experiencing reduced flow. This issue can be mitigated by installing pressure-compensating emitters, which maintain uniform flow rates regardless of elevation differences. These specialized emitters ensure that every vine receives consistent water delivery, regardless of its location within the vineyard.

Soil variability also poses a challenge, as different soil types have varying water retention and drainage capabilities. To address this, vineyard managers should adjust irrigation schedules based on soil moisture levels and the specific needs of the vines. Regular soil monitoring using sensors or moisture probes helps ensure that water is applied where it is needed most, avoiding over- or under-irrigation in different parts of the vineyard.

Emitter clogging is another common issue, often caused by debris, mineral deposits, or algae entering the irrigation system. Advanced filtration systems are essential for preventing these contaminants from reaching the emitters. Installing high-quality filters at the pump station and regularly maintaining them can significantly reduce the risk of clogging. In regions with hard water, chemical treatments to dissolve mineral buildup may also be necessary.

By implementing these mitigation strategies—pressure-compensating emitters, advanced filtration, and tailored irrigation schedules—vineyard managers can address the challenges of drip irrigation effectively. This proactive approach ensures the system operates efficiently, supporting healthy vine growth and optimizing grape production.

Drip irrigation is a highly efficient water delivery method widely used in vineyards. However, its effectiveness can be significantly hindered by plugging of emission devices. This issue arises from various factors, including sediment, organic matter, and chemical precipitates. Vineyard managers must understand the causes, treatment options, and preventive measures to maintain system performance and ensure uniform water distribution.

Several factors contribute to this issue, each posing unique threats to the system's efficiency and longevity.

Sediment and particulates are among the primary causes of plugging. Materials like sand, silt, or clay can infiltrate the irrigation system, often bypassing basic filtration units. These fine particles accumulate within the emitters, leading to blockages that restrict water flow. Without proper filtration and regular system maintenance, sediment buildup can become a persistent problem.

Organic material also poses a significant challenge, particularly in systems with high levels of clay particles in the irrigation water. Slimy bacteria thrive on these particles, forming biofilms along the interior walls of drip hoses and emitters. Over time, these biofilms can dislodge, traveling downstream and clogging emitters. This not only reduces system efficiency but also increases the need for chemical treatments and flushing protocols.

Chemical precipitates, including calcium carbonate, magnesium carbonate, and iron deposits, frequently occur in systems using hard water or those with inadequate pH management. These inorganic compounds crystallize and form deposits inside emitters, gradually reducing water flow. Such issues are particularly common when water sources contain high levels of dissolved minerals, necessitating the use of acid treatments or other chemical interventions to dissolve the deposits.

Nutrient application can exacerbate plugging issues, especially when organic fertilizers with high percentages of suspended solids are used. These fertilizers often leave residues that accumulate in emitters, creating additional blockages. Without careful management and system maintenance, the buildup from nutrient applications can significantly impair the overall performance of the drip irrigation system.

Addressing these causes requires a comprehensive approach that includes advanced filtration, chemical treatments, and routine flushing to ensure the system remains operational and efficient. Understanding the specific plugging risks associated with sediment, organic material, chemical precipitates, and nutrient applications is important for implementing effective prevention and maintenance strategies.

Filtration systems serve as the first line of defence against plugging. Properly installed and maintained filters are vital for removing particulates and debris from the irrigation water. Depending on water quality, filtration options may include tubular screens, disc filters, or media tanks. In cases where water sources contain high sand content, installing sand separators upstream enhances filtration efficiency by preventing abrasive particles from entering the system. Regular filter maintenance is key to ensuring continued effectiveness.

Chemical treatments play a critical role in preventing and resolving plugging issues caused by biofilms, carbonate deposits, and mineral precipitates. Acid treatments effectively dissolve carbonate deposits and lower water pH, preventing the future formation of insoluble compounds. Chlorine is widely used to inhibit biofilm growth, oxidize organic sediments, and improve overall system hygiene. For water sources with high organic loads, ozone can be used to eliminate organic material, although its short half-life limits its effectiveness at system extremities. Specialized polymers offer an additional layer of protection by sequestering iron and manganese, keeping these minerals in solution and preventing them from precipitating.

Flushing is a practical and essential maintenance practice for removing accumulated sediment and contaminants from the irrigation system. Regular flushing of mainlines, sub-mainlines, and drip hoses ensures that debris does not settle and obstruct water flow. High flushing velocities are required for optimal effectiveness, with recommended speeds of 1.5 m/s for mainlines and sub-mainlines and 0.5 m/s for drip hoses. These velocities help dislodge materials clinging to the interior surfaces of hoses and emitters, expelling them from the system.

Adjusted fertilizer injection is another important strategy to reduce the risk of plugging. By injecting fertilizers upstream of filtration units, any debris or residues introduced during the process are captured before they reach the emitters. This minimizes the likelihood of precipitate formation and ensures that nutrient delivery does not compromise the integrity of the irrigation system.

By combining these treatment options—effective filtration, targeted chemical treatments, consistent flushing, and optimized fertilizer injection—vineyard managers can prevent plugging, maintain high distribution uniformity, and ensure the efficient operation of their drip irrigation systems.

Preventing plugging and maintaining the efficiency of drip irrigation systems require proactive measures that focus on routine maintenance, strategic treatments, and optimized system design. These steps ensure consistent water delivery, reduce the risk of clogging, and extend the lifespan of the irrigation infrastructure.

Regular Maintenance is the cornerstone of a well-functioning irrigation system. Routine inspections and cleaning of critical components such as pumps, filters, and emitters are essential to prevent the buildup of debris and contaminants. Key parts like valves, pressure regulators, and flow meters should be periodically calibrated to ensure they are operating correctly and delivering accurate performance. Maintenance schedules should also include monitoring for leaks or damage that could compromise system efficiency.

Shock Treatments provide an effective solution for addressing severe plugging issues. In this process, a concentrated chemical dose is applied to clean the system thoroughly. Before injecting the chemical, hose ends are opened to flush out contaminants, ensuring that the treatment reaches all parts of the system uniformly. This method not only dislodges accumulated debris but also enhances the effectiveness of chemical treatments in clearing biofilms and deposits.

System Design Optimization is significant for minimizing the risks of uneven water distribution and contamination. Pressure-compensating emitters play a significant role by maintaining consistent flow rates across varying elevations, ensuring uniform water delivery throughout the vineyard. Additionally, the incorporation of well-designed filtration and chemical injection systems reduces the likelihood of debris and contaminants entering the irrigation lines. Properly positioned filters and injectors further safeguard the system against clogging.

Proper System Flushing is a critical maintenance practice to remove sediment and contaminants from drip hoses and pipelines. Flushing should continue until water draining from the hoses runs clear for at least two minutes. Vineyard managers should monitor for two distinct waves of contaminants during flushing. The first wave typically consists of solid materials collected at downstream hose ends, while the second wave results from the breakdown of sedimentation on hose interiors. Ensuring thorough flushing minimizes the risk of buildup and maintains the system's efficiency.

The flushing duration depends on the length of the pipe and the flow rate. The formula for calculating flushing time is:

$$\text{Flushing Time (minutes)} = \frac{\text{Length of Pipe (meters)} \times 60}{\text{Flow Rate (m/s)}}$$

Example Calculation:

- Pipe length: 100 meters

- Flow rate: 0.5 m/s

$$\text{Flushing Time} = \frac{100 \times 60}{0.5} = 12,000 \, \text{seconds} = 20 \, \text{minutes}$$

This ensures thorough removal of contaminants from the system.

Renewable Energy Options

Incorporating renewable energy into small-scale vineyards not only reduces environmental impact but also enhances economic sustainability by lowering operational costs over time. Renewable energy sources can power essential vineyard operations such as irrigation, temperature control, lighting, and equipment, aligning with the goals of sustainable viticulture.

Solar energy has emerged as a versatile and increasingly popular renewable energy solution for vineyards, primarily due to its adaptability and declining costs. The integration of solar panels into vineyard operations can take various forms, including rooftop installations, ground-mounted systems, and innovative agrivoltaic systems that allow for the simultaneous cultivation of crops and the generation of solar energy. Agrivoltaics not only optimize land use but also enhance the sustainability of vineyard practices by reducing the carbon footprint associated with traditional energy sources [99, 204].

Figure 16: Oregon vineyard in the Willamette Valley wine region utilizing solar power. eyeliam, CC BY 2.0, via Wikimedia Commons.

One of the significant applications of solar energy in vineyards is in irrigation systems. Solar-powered water pumps provide an energy-efficient alternative to conventional grid electricity or fossil fuel-powered pumps. This technology is particularly beneficial in regions where water demand peaks during sunny days, as solar irrigation systems can deliver water directly to crops when it is most needed [205, 206]. Studies have shown that solar-powered irrigation can be more economical than diesel systems, allowing for broader coverage and reduced operational costs [206]. Furthermore, the implementation of solar energy in irrigation supports sustainable agricultural practices and contributes to significant reductions in greenhouse gas emissions [206].

In addition to irrigation, solar energy plays a major role in cold storage and processing within the wine industry. Solar panels can effectively power cooling systems necessary for grape storage and small-scale wine processing equipment, ensuring that the quality of the product is maintained without relying on traditional energy sources [207]. The ability to harness solar energy for these applications enhances operational efficiency and aligns with the industry's growing emphasis on sustainability [207].

Moreover, solar energy systems can provide lighting and monitoring solutions for vineyards. These systems can power nighttime lighting and various sensors or cameras used for vineyard monitoring, thereby improving operational oversight and data collection [208]. The integration of solar energy into these aspects of vineyard management reduces reliance on grid power and minimizes operational costs and environmental impacts [208].

Figure 17: Close up of solar panels in the vineyard, taken at Left Coast Wines in the Eola-Amity Hills AVA of Oregon. Agne27, CC BY-SA 3.0, via Wikimedia Commons.

Despite the numerous advantages, the adoption of solar energy in vineyards is not without challenges. The initial setup costs can be substantial, and the effectiveness of solar systems is highly dependent on regional climate conditions, particularly sunlight availability [99, 209]. However, the long lifespan of solar installations, typically requiring minimal maintenance over 20-25 years, makes them a worthwhile investment for many vineyard operators [99, 204].

Wind energy has emerged as a vital component of the renewable energy landscape, particularly for small-scale applications such as those found in vineyards located in windy areas. The integration of small-scale wind turbines can provide significant benefits, including powering irrigation systems and serving as a supplementary energy source to solar power systems. This synergy is particularly advantageous in regions with variable weather patterns, where the intermittent nature of solar energy can be complemented by the consistent generation of wind energy.

Figure 18: Small scale wind turbine in a vineyard setting. Mitchell Henderson, CC0, via Pexels.

Small-scale wind turbines (SSWTs) are particularly well-suited for off-grid applications, such as powering water pumps for irrigation systems. These turbines can generate sufficient energy to operate pumps, ensuring that vineyards have a reliable water supply without relying on grid electricity, which may be unavailable in remote areas [210]. Furthermore, the combination of wind and solar energy systems enhances energy reliability. During periods of low solar generation, such as cloudy days or nighttime, wind energy can provide a backup, ensuring continuous power supply for vineyard operations [211, 212].

The advantages of wind energy are manifold. It is a renewable resource that contributes to reducing reliance on fossil fuels, thus mitigating environmental impacts [213]. Additionally, the operational costs of small-scale wind turbines are relatively low, making them an economically viable option for many vineyard owners [214]. The ability to harness wind energy in conjunction with solar power systems allows for a more resilient energy infrastructure, particularly in regions where energy demand fluctuates significantly [215].

Wind turbines require consistent wind speeds to operate efficiently, which may limit their applicability in certain locations [216]. Additionally, concerns regarding noise and visual impact can deter some vineyard owners from adopting this technology [217]. It is essential for potential users to conduct thorough site assessments to determine the feasibility of wind energy installations, taking into account local wind patterns and community acceptance [218].

Biomass energy represents a significant avenue for converting organic waste into usable energy, particularly in agricultural settings such as vineyards. The process involves transforming materials like pruned vines and grape pomace into energy through methods such as combustion and anaerobic digestion. This conversion not only addresses waste management but also contributes to renewable energy production, making it a sustainable option for energy generation.

Heating applications of biomass are particularly relevant in vineyard operations. Biomass can be utilized to heat greenhouses, fermentation tanks, and winery facilities, thereby enhancing operational efficiency and reducing reliance on fossil fuels [219, 220]. The ability to use local agricultural residues, such as vineyard pruning, for heating purposes exemplifies the closed-loop system that biomass energy can create, where waste is repurposed into energy, thus minimizing disposal costs and environmental impact [220, 221].

In addition to heating, biomass can also be harnessed for electricity generation. Biomass digesters can produce biogas, which can then be used to generate electricity for vineyard operations [219]. This dual functionality of biomass—providing both heat and electricity—highlights its versatility as a renewable energy source. Furthermore, the integration of biomass into energy systems can significantly reduce greenhouse gas emissions, as it offers a cleaner alternative to traditional fossil fuels [222, 223].

Despite its advantages, the implementation of biomass energy systems is not without challenges. Significant infrastructure investment is required to establish biomass conversion facilities, and regular maintenance is necessary to ensure efficient operation [224]. The efficiency of converting waste to energy is highly dependent on the scale of operation and the technology employed [225, 226]. For instance, the effectiveness of biomass systems can vary based on the type of biomass used and the conversion technology applied, such as gasification or pyrolysis [227, 228].

Moreover, logistical challenges related to the collection, storage, and transport of biomass can hinder its adoption as a mainstream energy source [225]. Developing densification technologies and improving transportation logistics are essential steps toward enhancing the viability of biomass as a renewable energy resource [225, 229].

Geothermal energy represents a sustainable solution for heating and cooling applications by utilizing the stable temperatures found underground. This technology is particularly beneficial in various sectors, including agriculture, where it can optimize conditions for activities such as winemaking and crop protection.

One of the primary applications of geothermal energy is through geothermal heat pumps (GHPs), which can effectively regulate temperatures in cellars or storage areas. These systems leverage the consistent underground temperatures to maintain optimal conditions for winemaking, ensuring that the environment remains stable regardless of external weather fluctuations [230, 231]. Furthermore, in colder climates, geothermal systems can provide frost protection for vineyards by warming the soil or air around the vines, thus safeguarding them against frost damage [232, 233].

The advantages of geothermal systems are significant. They are known for their high efficiency and low operating costs once installed, making them an economically viable option in the long term. For instance, ground source heat pump systems can lead to substantial savings in annual electricity consumption, often reported to be around 40% compared to conventional heating systems [234]. Additionally, the direct utilization of geothermal energy for heating and cooling applications, such as in district heating systems, has been widely adopted across numerous countries, showcasing its effectiveness and reliability [235, 236].

However, the implementation of geothermal systems is not without challenges. The initial installation costs can be considerable, which may deter potential users despite the long-term savings [237]. Moreover, the feasibility of geothermal energy systems heavily depends on local geological conditions, which can vary significantly from one region to another. This variability can affect the efficiency and effectiveness of the geothermal heat pumps, necessitating careful site assessments before installation [238, 239].

Hydropower represents a viable option for vineyards located near flowing water, leveraging the kinetic energy of streams or rivers to generate electricity. This small-scale hydropower can effectively power essential equipment such as irrigation pumps and vineyard machinery, providing a consistent energy source that is particularly beneficial in agricultural settings. The integration of hydropower systems into vineyard operations not only enhances energy reliability but also contributes to the sustainability of agricultural practices by utilizing renewable resources [240, 241].

The eco-friendly integration of hydropower systems is fundamental to minimizing ecological disruption while maximizing energy production. Properly designed hydropower systems can coexist with the natural environment, ensuring that the ecological balance is maintained. Research indicates that small hydropower technologies are particularly well-suited for rural electrification and can significantly reduce reliance on fossil fuels, thereby lowering greenhouse gas emissions associated with traditional energy sources [242, 243]. Furthermore, the use of hydropower aligns with the broader goals of sustainable agriculture, where renewable energy sources are increasingly recognized as essential for reducing the carbon footprint of farming practices [244].

However, the implementation of hydropower systems in vineyards is not without challenges. The availability of water resources is a critical factor that can limit the feasibility of such projects. In regions where water flow is inconsistent or subject to seasonal variations, the reliability of hydropower generation may be compromised [245, 246]. Additionally, regulatory approvals are often necessary to install hydropower systems, which can introduce delays and complexities in project development. Understanding the regulatory landscape and ensuring compliance with environmental standards is essential for the successful deployment of hydropower in agricultural contexts [240, 241].

The choice of renewable energy system depends on factors such as vineyard size, regional climate, available resources, and budget. A combination of systems, such as solar and wind or solar and biomass, often provides the most reliable and efficient solution. By adopting renewable energy technologies, small-scale vineyards can enhance sustainability, reduce costs, and contribute to a greener future for viticulture.

Chapter 4
Establishing the Vineyard

Preparing the Soil: Organic and Sustainable Techniques

Soil is increasingly recognized as a critical non-renewable asset that must be managed responsibly to ensure long-term agricultural productivity and environmental health. In Australia, this recognition is formalized through the State of the Environment reports produced every five years, which assess soil management practices among other environmental factors. In the context of viticulture, soil health plays a central role in supporting sustainable vineyard operations. Understanding and maintaining soil health involves evaluating its physical, chemical, and biological properties to ensure that it supports vine growth, enhances drought resilience, and reduces dependency on external inputs.

Soil health refers to the capacity of soil to function as a vital, living system that sustains biological productivity, maintains environmental quality, and promotes the health of plants and animals. It is a dynamic measure, influenced by farming practices, and reflects how well soil can meet the specific needs of its intended use. From a viticultural perspective, soil health is integral to the sustainability of farming practices and the long-term viability of vineyards.

Healthy soil provides a stable environment for grapevine roots, supporting vine longevity, consistent yields, and high wine quality. A well-managed soil system also enhances water infiltration, improves efficiency in water use, minimizes erosion, and recycles nutrients, thus reducing the need for chemical inputs. It creates conditions that promote natural biological processes and protects against compaction and off-site environmental impacts.

Soil is a three-phase system composed of solid matter, air, and water. Its physical structure is defined by the arrangement of these components, with solid particles forming aggregates and the spaces between them acting as pores. Macropores facilitate water drainage and air circulation, while micropores retain moisture for plant use. Soil porosity, often measured through bulk density, is a key indicator of soil health, with lower bulk density suggesting higher porosity and better aeration.

For vineyards, an ideal soil structure allows for deep root penetration and unrestricted growth. Grapevine roots can extend more than 60 cm in irrigated vineyards and up to 2 meters in non-irrigated conditions, with some roots reaching depths of 6 meters in gravely soils. Compaction layers, whether natural or caused by heavy machinery, can restrict root growth and water infiltration, negatively impacting vine health. Effective soil structure retains water in the root zone while ensuring adequate aeration following irrigation or rainfall.

The chemical composition of soil directly affects nutrient availability for grapevines. Healthy soil provides an adequate supply of macro- and micro-nutrients without excesses that could lead to toxicity or environmental pollution. Nutrients in the soil are often present as positively charged cations, which are retained by the negatively charged particles in clay and organic matter. The cation exchange capacity (CEC) of soil reflects its ability to retain and supply nutrients to plants.

Soil pH is a critical factor influencing nutrient availability. The ideal pH range for vineyard soils is 5.5 to 8, although many Australian vineyard soils naturally exceed this range in subsoil layers. Monitoring and adjusting soil pH, as well as nutrient levels, is essential for maintaining optimal vine nutrition. Soil tests, along with petiole and leaf blade analyses, help diagnose nutrient deficiencies or toxicities and guide appropriate remediation strategies.

Soil biology encompasses a diverse range of organisms, including fungi, bacteria, nematodes, worms, and insects, which play essential roles in nutrient cycling and soil structure. These organisms decompose organic matter, form humus, release carbon dioxide to break down minerals, and convert nutrients into forms accessible to plant roots. They also produce substances that stabilize soil aggregates, improving water retention and aeration.

A particularly important biological interaction in vineyards is the mycorrhizal association between grapevine roots and certain fungi. This symbiotic relationship enhances the vine's access to phosphorus, zinc, and water reserves, particularly in areas beyond the reach of root hairs. These beneficial interactions underline the importance of maintaining a diverse and active microbial community in vineyard soils.

Soil organic matter, although a small fraction of the soil's solid content, is vital for soil health. It consists of decomposed plant and animal material, including root exudates, and eventually forms humus. Organic matter enhances soil structure, water holding capacity, cation exchange capacity, and nutrient availability. It also buffers soil pH and provides a carbon food source for soil microbes. In vineyards, maintaining an organic matter content of 2-4% is ideal.

Improving soil health in vineyards involves practices that enhance organic matter content and support a thriving microbial ecosystem. Techniques include the application of composts, mulches, and cover crops. Composts, such as biochar or composted grape marc, improve long-term soil water retention and nutrient availability. Mulches reduce evaporation, suppress weeds, and moderate soil temperature, benefiting shallow feeder roots and promoting mycorrhizal associations. Cover crops prevent erosion, enhance biodiversity, and influence nitrogen levels depending on the crop type.

Soil health can be assessed through various tests measuring pH, nutrient levels, organic matter content, aggregate stability, and bulk density. While some tests can be conducted on-site, others require laboratory analysis. Research continues to explore the complex relationships between soil organisms and soil health, aiming to develop sustainable practices that foster beneficial microbial communities.

Soil preparation is a critical step in establishing a vineyard, directly influencing vine health, grape quality, and the long-term sustainability of viticulture. Organic and sustainable techniques focus on preserving the environment, improving soil health, and fostering ecosystem balance. By minimizing reliance on synthetic inputs, these practices create a resilient vineyard system capable of producing premium grapes with minimal environmental impact.

A comprehensive soil analysis is the first step in organic and sustainable soil preparation. Understanding the soil's composition, pH levels, organic matter content, nutrient profile, and drainage capacity allows for tailored interventions. Organic approaches avoid synthetic testing chemicals, instead relying on natural indicators such as earthworm activity and microbial diversity. Sustainable practices use soil test results to guide targeted amendments and management strategies that support natural fertility and balance.

Incorporating organic matter is fundamental for improving soil fertility, structure, and water retention. Compost made from vineyard waste, such as grape pomace, returns nutrients to the soil while supporting microbial activity. Aged manure from livestock raised on organic feed provides essential nitrogen, while green manures like clover or alfalfa are plowed into the soil to enhance organic content and nitrogen fixation. These inputs create a nutrient-rich environment that promotes healthy root growth and soil vitality.

Traditional tilling practices often disrupt soil structure, leading to erosion and loss of organic matter. Organic and sustainable methods advocate for no-till or reduced-till farming to preserve soil integrity and microbial habitats. No-till farming involves planting directly into crop residues or organic mulch without disturbing the soil. Reduced-till approaches limit tilling to shallow depths for planting rows while leaving surrounding soil intact. These practices maintain soil structure, reduce carbon emissions, and protect against erosion.

Cover cropping between vineyard rows offers numerous benefits, including erosion control, weed suppression, and biodiversity enhancement. Organic cover crops such as legumes fix nitrogen in the soil, while grasses like rye stabilize the soil. Rotating cover crop species ensures soil fertility and breaks pest cycles. Additionally, cover crops attract beneficial insects and pollinators, contributing to a balanced vineyard ecosystem.

Healthy soil microbiomes are essential for nutrient cycling, disease suppression, and overall vitality. Organic and sustainable methods prioritize enhancing microbial life through biofertilizers and compost teas. Natural inoculants such as mycorrhizal fungi form symbiotic relationships with vine roots, improving nutrient uptake and water access. Avoiding synthetic fungicides, herbicides, and pesticides protects these beneficial microorganisms, ensuring long-term soil health.

Managing soil pH organically is important for maintaining nutrient availability and vine performance. Lime from natural sources raises pH in acidic soils, while elemental sulphur or organic mulch lowers pH in alkaline conditions. These adjustments are made gradually and monitored regularly to avoid overcorrection, ensuring optimal conditions for vine growth.

Effective weed control during soil preparation reduces competition for nutrients and water. Organic mulches like straw or wood chips suppress weeds while conserving moisture. Dense-growing cover crops outcompete weeds naturally, and manual removal methods maintain soil integrity without synthetic herbicides. These techniques support a clean and nutrient-rich environment for grapevines.

Proper drainage is essential to prevent waterlogging, which can damage roots and reduce soil oxygen levels. Sustainable drainage solutions include incorporating sand or gravel into heavy clay soils and creating swales or contouring to guide water flow. Deep-rooted cover crops such as radishes break up compacted layers, improving water infiltration and soil aeration.

Erosion control is vital for protecting soil and the surrounding environment. Terracing on slopes prevents runoff, while grass strips or ground covers stabilize soil during rainstorms. Native vegetation around the vineyard serves as a buffer against wind erosion and helps stabilize the soil further. These measures safeguard the vineyard while enhancing its ecological resilience.

Reducing carbon footprints aligns with the sustainability goals of organic viticulture. Using renewable energy sources like solar or biofuels to power equipment minimizes emissions. Sourcing amendments locally reduces transportation-related carbon output, and employing low-impact machinery designed for small-scale farming conserves fuel. These practices ensure environmentally responsible vineyard operations.

By adopting organic and sustainable soil preparation techniques, vineyard managers create a fertile foundation for grapevines. These methods not only enhance soil health and vine productivity but also support a thriving ecosystem, ensuring the long-term success of the vineyard and its surrounding environment.

Organic Grape Growing Techniques

Over the past decade, the rise of gourmet cooking and healthy lifestyles has increased interest in organic produce and wine. Organic products, including wine, have carved a niche in supermarkets, produce stands, and farmers' markets across North America. This surge has sparked debates on whether organic farming truly enhances flavour and health benefits or if it's merely a trendy label catering to an "earthy" audience seeking to distance themselves from conventional agricultural practices. Ultimately, the quality of your wine will be the judge of this debate. Organic farming can be broadly defined as agriculture without synthetic fertilizers or pesticides, relying on diversity and natural competition to maintain pest control and ecological balance. It emphasizes the use of decomposed organic matter to nourish plants, creating a system that supports a balanced crop [247].

Backyard vineyards are particularly well-suited for organic wine grape production. Grapevines, which often require minimal fertilizers and chemical inputs, thrive in the United States and beyond with basic care. The challenges faced by organic vineyardists, such as pest control, disease management, soil fertility, weed suppression, and canopy management, are easier to address in small-scale vineyards. With a manageable number of vines, growers can closely monitor their vineyard and quickly adapt to the needs of individual plants, allowing for a hands-on approach to organic viticulture [247].

The essence of organic grape production lies in fostering a thriving, self-sustaining ecosystem. Organic farming requires growers to feed the soil, not the plant, strengthening the environment that supports grapevines. The goal is to reduce reliance on synthetic inputs by building soil fertility through homemade compost, natural fertilizers, and the promotion of biodiversity. By prioritizing healthy soils, organic vineyards can sustain themselves with minimal outside interference while producing flavourful grapes [247].

Sustainable farming shares many principles with organic methods but often involves additional considerations. Sustainability focuses on maintaining the health and productivity of the land for future generations. This approach reduces erosion, soil depletion, and chemical dependency while promoting soil health and ecological harmony. Sustainable practices encourage growers to consider environmental, social, and economic viability, ensuring a holistic approach to vineyard management. For backyard growers, this may include using on-site compost, conserving water, and integrating renewable energy sources where possible [247].

A critical concept in organic and sustainable viticulture is biodiversity. A balanced ecosystem supports healthy vines by leveraging natural pest control, soil enrichment, and disease resistance. For example, maintaining a diverse habitat of animals and insects can naturally regulate pest populations, reducing the need for chemical interventions. Soil biodiversity is equally vital, as microorganisms break down organic matter into nutrients, stabilize soil structure, and combat soil-borne pathogens. This interconnectedness between above-ground and below-ground diversity ensures a robust and resilient vineyard system [247].

Transitioning to organic practices should be approached gradually. Abrupt changes can jeopardize grape quality, particularly if synthetic inputs have been integral to previous practices. A phased transition allows growers to refine their methods while ensuring their crop remains healthy and productive. Building soil fertility organically takes time but yields long-term benefits, such as improved root environments and natural nutrient cycling. Observing the vineyard regularly is essential for identifying issues early, allowing for proactive and less invasive solutions [247].

Organic farming often involves creative strategies to tackle vineyard challenges. For instance, composting lawn clippings and kitchen waste provides nutrient-rich amendments, while mulching vine rows conserves moisture and suppresses weeds. Leaf pulling and pruning improve air circulation, reducing mildew risks and enhancing fruit exposure. Addressing erosion through cover crops and introducing beneficial insects with native wildflowers are further examples of how organic techniques integrate natural processes to maintain vineyard health [247].

Ultimately, the goal of organic and sustainable viticulture is to grow high-quality wine grapes while promoting environmental stewardship. By striking a balance between effective vineyard practices and sustainable principles, growers can enjoy a rewarding, successful, and environmentally conscious approach to winemaking. The process may require more planning and observation than conventional methods, but the result is often healthier vines, better fruit, and wines that truly reflect the terroir in which they are grown [247].

Setting up an Organic Vineyard

Planning for the future conversion of a vineyard to organic production involves thoughtful infrastructure and management decisions that align with organic principles while allowing for a conventional start. This approach balances the need for a viable establishment phase with the long-term goal of organic certification, avoiding unnecessary costs and setbacks caused by replanting or ineffective early practices. By designing the vineyard infrastructure to suit organic production, growers can minimize transition challenges and expenses later on [248].

While the specific requirements for establishing an organic vineyard overlap with those of conventional methods—such as site selection, variety choice, trellis design, irrigation, and nutrient management—the primary distinction lies in the limited use of synthetic chemical inputs. Conventional inputs can be strategically used during the initial phases to ensure cost-effective establishment without compromising future organic certification. For example, applying non-residual knockdown herbicides before planting effectively manages weeds without leaving lasting residues. Similarly, 'soft' insecticides and materials acceptable in organic systems can address pests and diseases affecting young vines, ensuring their survival and early growth [248].

Fertilizers like superphosphate and calcium nitrate may be used during the establishment phase due to their readily available nutrient forms. Australian soils, which are often low in native phosphorus, benefit from the application of superphosphate, providing essential nutrients to support young vines. Calcium nitrate is a less harmful nitrogen source and can be applied through irrigation systems to sustain early vine growth. Monitoring through soil and tissue testing ensures fertilizers are applied at appropriate rates, avoiding over-application that could hinder future organic compliance [248].

Soil preparation is a critical step for successful vineyard establishment and should begin a year before planting. Fertilizing the soil, implementing effective weed control measures, and potentially establishing cover crops create a fertile and balanced growing environment. Nutrient management in organic systems might involve using compost, rock phosphate, or guano as amendments. Biodynamic growers may incorporate biodynamic preparations to enhance soil vitality. Deep ripping of the soil in multiple directions is a standard practice for both conventional and organic setups, breaking up hardpans and improving root penetration. Organic growers, however, replace synthetic options like superphosphate with untreated rock phosphate, recognizing that the latter releases nutrients more slowly, especially in alkaline soils [248].

Lime is often required to adjust soil pH in acidic regions, a practice acceptable in organic vineyards if the lime is untreated and sourced naturally. Compost and mulch applications around young vines serve dual purposes, acting as fertilizers and conserving soil moisture. While beneficial, care must be taken during compost breakdown to ensure nitrogen availability is not adversely affected. These materials are often applied using side delivery systems, particularly in larger-scale operations [248].

A practical example of this process includes spreading gypsum, grape marc, and chicken manure as soil amendments, followed by deep cross-ripping to improve soil structure. Installing posts and a drip irrigation system creates a robust framework for vine establishment. In the early stages, herbicides like Roundup may be used judiciously for weed control, ensuring a clean start for planting lines. By the third year, growers can expect sufficient vine development to transition into the organic conversion phase, leveraging the infrastructure established during the conventional phase [248].

For those starting entirely within organic or biodynamic principles, the focus remains on soil preparation, vineyard layout, irrigation design, and planting stock selection. Pre-planting activities might include applying rock phosphate, guano, or compost to cover crops and using cultivation for initial weed control. Organic soil amendments, such as untreated lime, compost, and mulch, improve soil fertility and moisture retention while adhering to organic standards. Compost breakdown requires careful management to prevent nitrogen imbalances that could impact young vines. Deep ripping along planting lines ensures optimal root growth, while the use of untreated natural materials supports sustainable practices [248].

By integrating these strategies, growers can lay a strong foundation for their vineyards, ensuring a seamless transition to organic certification while fostering sustainable and productive vine growth. This forward-thinking approach emphasizes soil health, biodiversity, and environmentally responsible practices, aligning with the broader goals of organic viticulture [248].

Planting and Trellising Systems

Grapevine trellis systems are essential in viticulture, serving as the structural backbone for supporting grapevines in a controlled and organized manner. These systems play a vital role in optimizing sunlight exposure, enhancing air circulation, and ensuring accessibility to fruit. Each trellis system is designed to address specific vineyard conditions, such as grape variety, topography, climate, and management practices, directly influencing vine health, yield, and grape quality.

Figure 19: Riversdale Estate Vineyard - vines under 'lyre trellis'. Riversdale Estate, CC BY 2.0, via Flickr.

Trellis systems fulfill several critical purposes. They provide the structural support needed to hold the weight of vines, shoots, and grape clusters, preventing them from sagging or touching the ground. By positioning foliage optimally, trellises maximize sunlight capture, which is essential for photosynthesis and the ripening of fruit. They also improve air circulation by spacing out the canopy, reducing humidity and the risk of fungal diseases such as mildew. Additionally, trellis systems simplify the management of pruning, canopy adjustments, pest control, and harvesting while guiding vine growth to prevent overcrowding and make efficient use of vineyard space.

One of the simplest trellis systems is the single-wire trellis. It features a single horizontal wire to support the vine, allowing shoots to grow naturally upward or be manually trained. This system is suitable for small-scale vineyards or less vigorous varieties, particularly in regions with low rainfall. While it is cost-effective and easy to install, it has limited options for canopy management and may not suit highly vigorous varieties. A more common system is vertical shoot positioning (VSP), which utilizes multiple wires to support shoots trained upward, positioning the fruiting zone below the foliage. This system is ideal for regions with ample sunlight and moderate vine vigour, offering benefits such as improved sunlight exposure, airflow, and compatibility with mechanical pruning and harvesting. However, it may not perform well in regions with excessive rainfall or for vigorous varieties.

For managing highly vigorous vines in fertile soils, systems like the Geneva Double Curtain (GDC) and the T-Trellis are effective. The GDC uses two horizontally positioned fruiting wires on either side of the row, with shoots trained downward. This design reduces canopy density, improves airflow, and increases sunlight penetration, supporting high yields but requiring more space and labour-intensive management. Similarly, the T-Trellis creates two canopy zones on either side of a single horizontal bar and encourages balanced growth and reduced shading, though it also demands a more complex setup and maintenance.

In traditional viticulture, pergola or overhead trellis systems train vines to grow above head height, forming a canopy. These systems are ideal for table grapes or ornamental purposes and provide shade in areas with intense sun exposure. They offer aesthetic appeal and protection against sunburn but are labour-intensive and less effective for wine grape production. Another option for vigorous varieties is the Lyre system, which divides the canopy into two arms extending outward to improve ventilation and light exposure. However, it requires significant space and initial investment.

More specialized systems like the Scott Henry and Smart-Dyson methods feature vertically divided canopies. Scott Henry trains half the shoots upward and the other half downward, while Smart-Dyson alternates upward and downward shoots. These systems enhance light distribution, reduce shading, and balance vine growth but require precise management and additional labour. For traditional or low-resource vineyards, the bush or goblet system is an ancient approach without trellises, where vines are pruned into a goblet shape for self-support. While it is cost-effective and suitable for arid regions, it has limitations in productivity and canopy management.

Choosing the right trellis system depends on various factors, including vine vigour, climate, soil fertility, available labour, and vineyard goals. Vigorous varieties benefit from systems like GDC or T-Trellis, while systems like VSP are well-suited to sunny, dry climates. In fertile soils, divided canopies can manage dense foliage, and for mechanization, systems such as VSP or Scott Henry are ideal. A pergola system might be the choice for aesthetic or ornamental purposes.

A thoughtfully planned trellis system significantly enhances vineyard productivity and sustainability. By adapting vine growth to specific local conditions, these systems ensure high-quality grape production while simplifying management and promoting long-term vineyard health.

Figure 20: Trellised vines, Coriole Vineyard, McLaren Vale South Australia. Rexness from Melbourne, Australia, CC BY-SA 2.0, via Wikimedia Commons.

Applying Organic Principles

Designing a vineyard layout and selecting an appropriate trellis system are fundamental steps in establishing a productive and sustainable vineyard, particularly in organic viticulture. The layout and trellis design must consider environmental factors, vine health, and long-term management needs to optimize grape production while adhering to organic principles.

Organic practices emphasize planting vines along contour lines to enhance water infiltration and minimize erosion and runoff. This approach helps preserve soil structure and reduces the environmental impact of vineyard operations. The layout must also be carefully planned to maximize air drainage and light availability within the canopy. Proper airflow reduces humidity levels, thereby lowering the risk of fungal diseases, while sufficient light exposure is essential for photosynthesis and fruit development. Close planting of vines is discouraged if it leads to increased shading and reduced air circulation, as this can compromise vine health and productivity.

The choice of trellis system significantly influences air movement, light penetration, and overall vine performance. Open systems, such as the Lyre trellis or Smart-Dyson system, are often favoured for their ability to enhance airflow and light availability throughout the canopy. These systems are particularly beneficial in regions prone to high humidity and vigorous vine growth, as they help mitigate the risk of fungal diseases. However, open trellis systems like the Lyre are more expensive to install and are incompatible with mechanical harvesting, limiting their application in large-scale or mechanized vineyards.

Vertical shoot positioning (VSP) systems are another effective option, particularly in organic vineyards where vine vigour is typically lower than in conventional operations. VSP systems are well-suited to organic practices due to their simplicity and compatibility with lower input levels. However, in high-vigour sites, vertical systems may result in shading of basal buds, which can reduce fruit set for the following year. This cycle of reduced fruitfulness must be carefully managed through appropriate pruning and canopy management strategies. The choice of trellis system should also consider local climatic and viticultural conditions, as the importance of light penetration varies across regions. In hot, dry climates, trellis design plays a less critical role compared to cooler, high-rainfall areas where vigorous growth and high humidity necessitate open systems to manage disease pressure.

Training vines to align with the chosen trellis system typically begins in the second year of growth. Organic vineyards often exhibit lower vine vigour due to reduced inputs of synthetic fertilizers and irrigation compared to conventional vineyards. Consequently, training demands may occur later in organic systems, allowing for a more gradual establishment phase. In contrast, conventional vineyards with higher vigour may require earlier attention to training and canopy management to control excessive growth.

Vine spacing is another critical consideration influenced by factors such as climate, soil type, expected vine vigour, variety, and available equipment. In low-vigour sites, closer spacing may be advantageous to maximize land use without causing overcrowding. Conversely, high-vigour sites may require wider spacing to prevent competition between vines and ensure adequate air and light penetration. There are no universal standards for vine spacing, with densities ranging from 1,500 vines per hectare in low-density plantings to as high as 3,000 vines per hectare in high-density systems. The choice of spacing should reflect site-specific conditions and production goals, balancing the need for efficient land use with the maintenance of optimal vine health.

A well-designed vineyard layout and trellis system, tailored to the specific conditions of the site, form the foundation of a successful organic vineyard. By optimizing airflow, light availability, and spacing, vineyard managers can promote healthy vine growth, enhance grape quality, and ensure long-term sustainability.

Trellis Construction

Building a vineyard trellis is a critical step in establishing a productive and profitable vineyard. As one of the most significant expenses during the vineyard's initial setup, the trellis structure directly impacts the long-term health, yield, and financial success of the vineyard. Proper planning and material selection are essential to ensure durability, functionality, and compatibility with the chosen vineyard management system.

The trellis system serves as the backbone of vine support, requiring strong and well-anchored end posts that can handle the tension of high-tensile wires and the weight of the vines as they grow and bear fruit. End posts are typically constructed using heavy wood or metal and are secured with robust braces to maintain stability under stress from wind, fruit load, and mechanical operations. The interior line posts, which provide additional support along the vine rows, are generally smaller in diameter and spaced at regular intervals. Constructing the trellis during the first growing season is vital to prevent damage to young vines from construction activities or equipment.

Line Posts and Materials: Line posts are a major cost component of trellis construction. The choice of materials depends on availability, cost, equipment for installation, and personal preference. Wooden posts made from rot-resistant species like black locust or cedar can last over 20 years, while untreated wood from less durable species may last less than a decade. Pressure-treated wooden posts, injected with preservatives, are widely used due to their enhanced longevity and structural reliability. Metal posts are another popular option, offering ease of handling and installation. They can be competitively priced compared to wood and are often preferred for their durability and low maintenance requirements.

Trellis Wire: Trellis wires play a dual role in supporting the weight of the vines and orienting shoots for optimal growth. High-tensile galvanized steel wire is commonly used for its superior strength and durability, with tensile strengths exceeding 200,000 pounds per square inch. Proper attachment of wires to end and line posts is crucial for maintaining wire tension and system integrity. Wires can be wrapped and twisted around end posts or secured with galvanized staples or drilled holes in the line posts. Staples should be driven deep enough to hold the wires securely but still allow for tension adjustments when necessary.

End Posts and Bracing Systems: The end posts and their anchoring systems are critical components of any trellis. They provide the foundation for the trellis, bearing the force exerted by tensioned wires, vine weight, and environmental stressors. Several end assembly designs are used, depending on vineyard requirements.

The inverted "V" end assembly is ideal for high trellises and uses posts angled away from the row to distribute stress effectively. This system provides stability and is anchored with steel screw anchors for added support. The H-brace assembly is considered the strongest design, with two vertical posts spaced apart and connected by a horizontal cross-member. This setup distributes forces across a larger area, minimizing strain on individual posts. The tie-back assembly transfers load through the end post to a ground anchor, offering another reliable method for securing the trellis system.

Mechanically Farmed Vineyards: For mechanized farming, simplicity in trellis and training system design is essential. Using fewer trellis configurations reduces the complexity of equipment requirements and minimizes the time needed for adjustments and maintenance. Mechanized vineyards benefit from standardized systems that streamline operations and reduce costs over time.

Cost and Economic Considerations: Trellis construction represents a significant portion of the initial investment in a vineyard. While cost-effective solutions may seem attractive, careful consideration of material quality and design characteristics is vital for ensuring long-term productivity and profitability. A well-built trellis supports efficient vine growth, facilitates management practices like pruning and harvesting, and withstands environmental pressures, ultimately contributing to the vineyard's success.

Choosing the Best Trellis for Your Grapes

Choosing a trellis system for a new vineyard or modifying an existing one is a critical decision that goes beyond economic factors. It requires a thorough understanding of the vineyard's unique characteristics, including environmental conditions, vine growth habits, vigour, and the degree of mechanization desired. Each vineyard presents distinct challenges and opportunities, making the selection process highly specific to its individual needs.

Environmental conditions play a significant role in determining the appropriate trellis system. Factors such as temperature, topography, soil type, rainfall, and wind influence vine growth and canopy development. In regions with warm summer temperatures and abundant sunlight, vines tend to develop large, dense canopies, which require trellis systems capable of providing ample support and maintaining adequate airflow. Conversely, in cooler climates or areas with consistent, high-velocity winds, vine growth is often less vigorous, allowing for simpler trellis designs. Soil texture and rooting depth further impact vine vigour; deep, fertile soils may promote robust growth requiring more expansive trellising systems, whereas shallower or less fertile soils may support lower-vigour vines.

The natural growth habit of the grape variety is another key consideration when choosing a trellis system. For instance, many American grape varieties and their hybrids exhibit procumbent growth habits, meaning their shoots tend to trail toward the ground. These varieties may require training systems that lift and orient the shoots away from the vineyard floor to optimize sunlight exposure and air circulation. In contrast, upright-growing varieties may be well-suited to vertical shoot positioning systems, which allow shoots to be trained upward along wires.

Vine vigour significantly influences the selection of trellis systems. Highly vigorous vines, which produce abundant growth and larger canopies, necessitate robust and expansive trellising solutions to support the additional foliage and fruit load. Multi-wired systems with adjustable foliage wires can effectively manage vigorous growth by creating defined canopy zones that improve light penetration and airflow. On the other hand, low-vigour vines, which generate less foliage, can thrive with simpler systems like single-wire trellises, reducing the complexity and cost of installation and maintenance.

Mechanization is increasingly important in modern viticulture, particularly for larger vineyards seeking to optimize labour efficiency and reduce costs. The choice of trellis system can significantly impact the extent to which mechanization can be implemented. While all trellis and training systems can accommodate some level of mechanization, certain designs lend themselves better to automated operations. For example, systems like vertical shoot positioning (VSP) are often preferred in mechanized vineyards because they simplify tasks such as pruning, canopy management, and harvesting. However, highly customized or complex systems, such as pergolas or divided canopies, may require additional equipment or adjustments, increasing operational costs and complexity.

In conclusion, selecting the right trellis system requires a careful balance of environmental conditions, vine characteristics, and operational goals. By tailoring the trellis design to the specific needs of the vineyard, growers can enhance vine health, optimize fruit quality, and ensure the long-term productivity and sustainability of their operations. The right choice supports efficient management practices and aligns with both the vineyard's natural attributes and its economic objectives.

Training Systems

Training systems used in vineyards are generally categorized into non-divided and divided canopy systems. These systems play a critical role in optimizing vine growth, fruit production, and quality by influencing the exposure of foliage to sunlight and managing the overall canopy microclimate. The choice of a training system depends on factors like vine vigour, soil fertility, climate, and vineyard management goals.

Non-divided canopy systems feature a single fruit zone, making them simpler and more cost-effective to establish and maintain compared to divided canopy systems. These systems are ideal for low- to moderate-vigour vines and are commonly used in cooler climates or areas with limited resources. One of the most prevalent non-divided systems is the Vertical Shoot Position (VSP). In this system, shoots are trained upward along wires, restricting the fruiting and renewal zone to a small vertical space. VSP is particularly effective in cool-climate regions, where it promotes shoot vigor and ensures sunlight exposure to buds and fruit. However, in high-vigour vines, the canopy may become overly dense, necessitating leaf removal to improve air circulation and reduce disease risk.

Another non-divided system is the Umbrella Kniffin, which uses two- or three-wire trellises to support long canes originating from renewal spurs. This method, suited for moderate-vigour vines, positions the fruiting zone higher off the ground, improving air circulation and reducing disease potential. Similarly, the High Bilateral Cordon system involves training cordons along the top wire of a trellis, allowing for semipermanent extensions of the trunk. This system is simple to construct, typically requiring only two wires, and is effective for low- to moderate-vigour vines.

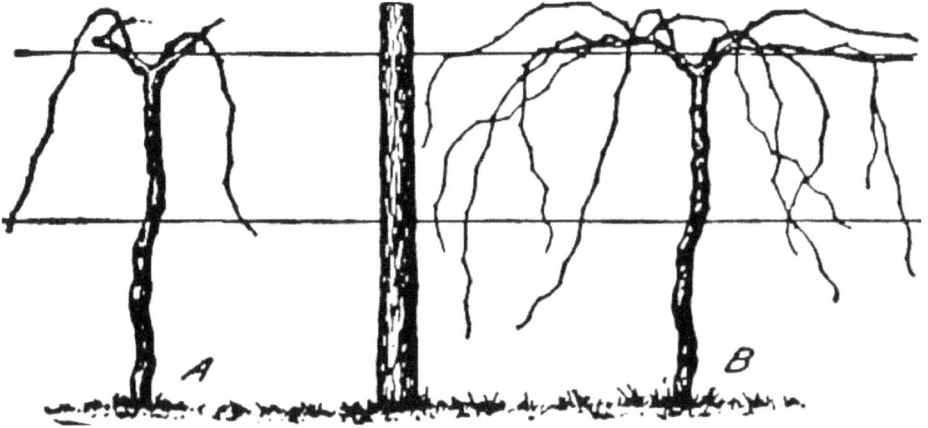

Figure 21: The umbrella kniffin system. Internet Archive Book Images, CC0 1.0, via Flickr.

Divided canopy systems, on the other hand, feature two fruit zones and are designed to expose more foliage to sunlight, thereby increasing yields while maintaining fruit quality. These systems are particularly suited for high-vigour vines or fertile sites where a single canopy would result in overcrowding. However, they are more complex and costly to establish.

One common horizontally divided system is the Geneva Double Curtain (GDC). The GDC divides the canopy into two parallel hanging curtains supported by a three-wire trellis. The system enables high bud counts and is well-suited for vigorous vines. The Lyre System, another horizontally divided option, is characterized by a U-shaped trellis structure. It is designed for upright-growing varieties and supports medium to high yields by maintaining equidistantly positioned spurs on lateral cordons.

Vertically divided systems include the Scott Henry and Smart-Dyson systems. The Scott Henry system separates the canopy into upward-growing (phototropic) and downward-growing (geotropic) shoots. This division balances vine growth and improves both yield and fruit quality, especially in high-capacity cool climate vineyards. The Smart-Dyson system operates similarly but trains shoots from the same cordon or fruiting zone, distinguishing it with its cordon-pruned structure. These systems are advantageous in areas where vertical canopy division can enhance light penetration and airflow, critical for disease prevention and fruit ripening.

In conclusion, the choice between non-divided and divided canopy systems depends on the specific requirements of the vineyard, including vine vigour, environmental conditions, and desired production goals. Non-divided systems are simpler and cost-effective for low-vigour sites, while divided canopy systems are suitable for vigorous vines, offering higher yields and better canopy management. Both approaches require careful consideration of vineyard characteristics to ensure optimal productivity and fruit quality.

Example Trellis Selections

Below are some real-life examples of small-scale sustainable wine-growing trellis selection choices, showcasing how growers have tailored their systems to their specific environmental, varietal, and sustainability goals:

Example 1: Vertical Shoot Positioning (VSP) in a Cool-Climate Vineyard

- **Location**: Oregon, USA

- **Grape Variety**: Pinot Noir

- **Site Characteristics**: Cool climate, moderate rainfall, sloping topography with well-drained sandy loam soil.

- **Trellis Choice**: VSP

- **Reasoning**:

    o The cool climate and moderate vine vigour of Pinot Noir are well-suited to VSP, which optimizes sunlight exposure and airflow in the canopy.

    o The trellis design facilitates efficient pruning, leaf pulling, and harvesting, all of which are essential for small-scale growers managing resources carefully.

    o The sloping terrain allows natural air drainage, further reducing disease pressure.

- **Sustainability Aspects**:
  - o Native cover crops are planted between rows to reduce erosion and promote biodiversity.
  - o Organic practices, including the use of compost teas and sulphur sprays, are integrated to manage pests and diseases naturally.

Example 2: Lyre Trellis in a Vigorous, High-Rainfall Area

- **Location**: Hunter Valley, Australia

- **Grape Variety**: Shiraz

- **Site Characteristics**: High rainfall, fertile soil with clay content, and high vine vigour.

- **Trellis Choice**: Lyre Trellis

- **Reasoning**:
  - o The divided canopy of the Lyre trellis reduces shading and increases airflow, minimizing the risk of fungal diseases in the humid environment.
  - o The system effectively manages the high vigour of Shiraz vines by spreading out the canopy and preventing overcrowding.
  - o While labour-intensive, the trellis ensures the quality of fruit by promoting even ripening.

- **Sustainability Aspects**:
  - o Chicken manure and composted grape marc are used as fertilizers to enhance organic matter and improve soil structure.
  - o Permanent cover crops between rows mitigate runoff and improve water infiltration.

Figure 22: This style of trellising is a type of Lyre spur training that creates essentially a "U-shape" with the vines, keeping the grapes high off the ground. John Morgan, CC BY 2.0, via Wikimedia Commons.

Example 3: Bush Vine System in an Arid Region

- **Location**: Santorini, Greece

- **Grape Variety**: Assyrtiko

- **Site Characteristics**: Hot, dry climate with volcanic soil and frequent high winds.

- **Trellis Choice**: Bush or Goblet System

- **Reasoning**:

  o The traditional bush vine system is well-adapted to the arid conditions, allowing the vines to shield their fruit from intense sun and drying winds.

- o The low-profile system minimizes water loss by reducing transpiration.

- o The design integrates seamlessly with the natural landscape, supporting cultural and ecological heritage.

- **Sustainability Aspects**:

  - o Mulching with volcanic ash improves water retention and minimizes evaporation.

  - o No irrigation is used, relying solely on natural rainfall to support vine growth.

Figure 23: Bush Training in Africa. Andrew Teubes, CC BY-SA 2.0, via Wikimedia Commons.

Example 4: T-Trellis in a Small Family Vineyard

- **Location**: Napa Valley, California

- **Grape Variety**: Cabernet Sauvignon

- **Site Characteristics**: Fertile valley floor with moderate rainfall and warm summers.

- **Trellis Choice**: T-Trellis

- **Reasoning**:

    o The T-Trellis allows for dual canopy zones, optimizing light penetration and air circulation for a high-yielding variety like Cabernet Sauvignon.

    o The system reduces labour for pruning and canopy management compared to more complex designs.

    o It supports sustainable goals by balancing vine vigour and improving yield quality.

- **Sustainability Aspects**:

    o Solar-powered irrigation systems are used to conserve energy.

    o Composting vineyard prunings and using them as mulch reduces waste and enhances soil fertility.

Example 5: Pergola System in a Historic Vineyard

- **Location**: Veneto, Italy

- **Grape Variety**: Garganega (used for Soave wine)

- **Site Characteristics**: Historical terraced vineyard with limestone soil and a mild Mediterranean climate.

- **Trellis Choice**: Pergola System

- **Reasoning**:

- o The pergola system aligns with the traditional practices of the region, preserving the vineyard's cultural and historical identity.

- o The overhead canopy protects grapes from sunburn and ensures even ripening in the Mediterranean climate.

- o It allows air to circulate freely, reducing disease risks.

- **Sustainability Aspects**:

  - o Drip irrigation is used sparingly to conserve water resources.

  - o Biodiversity is encouraged by maintaining wildflower borders and native vegetation around the vineyard.

Figure 24: Vineyard with wooden pergola. Milada Vigerova, Public Domain, via Pexels.

These examples highlight how trellis system selection can be adapted to the unique challenges and opportunities of each vineyard while maintaining a commitment to sustainable practices.

Sustainable Materials and Practices

Sustainable trellising for grapevines involves selecting materials and employing practices that minimize environmental impact while maintaining functionality, durability, and cost-effectiveness. A focus on sustainability in trellising not only benefits the environment but also contributes to the long-term viability and profitability of vineyards.

Using eco-friendly and renewable materials is at the core of sustainable trellising. These materials should be durable, locally sourced where possible, and require minimal processing.

Wood is a common choice for trellis posts, particularly when sourced from sustainably managed forests. Certified sustainable timber, such as that carrying Forest Stewardship Council (FSC) certification, ensures that the wood is harvested responsibly. Species like cedar or black locust are often preferred because they are naturally rot-resistant, reducing the need for chemical treatments.

Metal posts made from recycled steel or aluminium offer a sustainable alternative. These materials are durable, weather-resistant, and can be reused or recycled at the end of their lifespan. Metal posts often require less maintenance than wood, which makes them an appealing option for reducing long-term costs and environmental impact.

Composite posts made from recycled plastics and wood fibres are gaining popularity as a sustainable option. They are highly durable, resistant to rot and pests, and do not require chemical treatments. Their lightweight nature makes transportation and installation more efficient, further reducing carbon emissions.

High-tensile galvanized steel wire is a sustainable choice for trellising due to its strength and longevity. This wire reduces the need for frequent replacements and maintenance, lowering the overall material consumption over the vineyard's lifespan. Coated wires can also minimize environmental degradation caused by rust or corrosion.

Sustainable practices ensure that trellis construction and maintenance are efficient, eco-friendly, and supportive of vineyard ecosystems.

Building trellises with precise planning ensures minimal waste of materials. For instance, selecting appropriate post spacing and wire tension reduces overuse while maintaining structural integrity. Pre-fabricated trellis components can also reduce material waste and labour costs during construction.

When renovating or expanding a vineyard, reusing existing trellis materials, such as posts and wires, can significantly reduce resource consumption. Similarly, recycling old or damaged materials helps divert waste from landfills and reduces the demand for new raw materials.

Sourcing trellis components locally minimizes transportation-related carbon emissions and supports regional economies. Local materials are often better suited to withstand the specific environmental conditions of the vineyard's location.

Using natural or minimally invasive anchoring methods for trellis systems reduces soil disturbance and maintains ecosystem health. For example, screw-type anchors made from recycled metals can be installed without heavy machinery, preserving soil structure and reducing emissions from equipment.

Sustainable trellising practices can incorporate features that promote biodiversity, such as planting cover crops or native vegetation around trellis rows. These practices support beneficial insects and soil health, contributing to a more resilient vineyard ecosystem.

Sustainable trellising emphasizes long-lasting designs that require minimal maintenance. Trellis systems should be inspected regularly for damage, ensuring timely repairs that prevent the need for frequent replacements. Durable materials such as rot-resistant wood or galvanized steel can withstand harsh environmental conditions, reducing the environmental footprint over time.

Avoiding chemically treated materials, such as pressure-treated wood with harmful preservatives, minimizes the risk of leaching into the soil and surrounding ecosystem. Opting for natural or organically certified treatments ensures environmental safety.

Adopting sustainable materials and practices in grapevine trellising has numerous benefits. It reduces the carbon footprint of vineyard operations, preserves natural resources, and enhances the vineyard's overall ecological health. Additionally, these practices can improve the vineyard's marketability by appealing to environmentally conscious consumers and aligning with certifications for organic or sustainable viticulture.

Initial Watering and Fertilization

Initial watering is a critical practice in establishing vineyards, as it lays the groundwork for healthy vine growth and long-term productivity. This process is essential for ensuring that young vines receive adequate moisture, which is important for root establishment, nutrient uptake, and overall development. The significance of initial watering cannot be overstated, as it directly influences the survival rate and vigour of newly planted vines.

When grapevines are first planted, their root systems are typically underdeveloped, making it difficult for them to access moisture from deeper soil layers. Initial watering serves several key functions: it hydrates the root zone, reduces plant stress, facilitates nutrient uptake, and enhances soil settling around the roots. Hydration of the root zone is vital for promoting root extension and establishment, which is critical for the long-term health of the vines [249]. Furthermore, adequate watering minimizes water stress during the transplanting process, allowing vines to adapt more effectively to their new environment [250]. Moist soil conditions also enhance the solubility of nutrients, making them more readily available to the vines [251]. Additionally, watering helps to settle the soil around the roots, eliminating air pockets that can impede root-to-soil contact, which is essential for nutrient and water absorption [250].

The timing and frequency of initial watering are contingent upon various factors, including climate, soil type, and the specific needs of the vines. In hot, dry climates, more frequent watering may be necessary to prevent the soil from drying out, while cooler, wetter climates may require less frequent irrigation [252]. Soil type also plays a significant role; sandy soils, which drain quickly, may necessitate more frequent watering, whereas clay soils retain moisture and may require less frequent but deeper watering [252]. Young vines, with their smaller root systems, typically need more frequent watering compared to mature vines, which can access deeper moisture [249].

Several techniques can be employed for initial watering, each with its advantages and disadvantages. Drip irrigation is often considered the most effective method, as it delivers water directly to the base of each vine, ensuring precise application and minimizing waste [249]. Hand watering can be a viable option for smaller vineyards or areas without established irrigation systems, allowing for close monitoring of each vine's needs, albeit with increased labour intensity [249]. Flood irrigation may be suitable in specific contexts but can lead to inefficiencies and potential soil erosion if not managed carefully [249]. Overhead sprinklers provide broad coverage but may result in water loss through evaporation and increase humidity around the foliage, potentially fostering fungal diseases [249].

The volume of water required for initial watering is influenced by soil moisture levels, planting depth, and environmental conditions. Generally, the goal is to moisten the soil to a depth of 12–24 inches (30–60 cm) [250]. Sandy soils may require approximately 5–10 gallons (19–38 litres) of water per vine per session, while clay soils may need less water but longer watering durations to ensure deep penetration [250].

To optimize initial watering, several best practices should be followed. Monitoring soil moisture using tools like moisture probes can help determine the appropriate watering schedule [252]. It is necessary to avoid overwatering, as excess water can lead to waterlogging, suffocating the roots and promoting fungal diseases [249]. Applying organic mulch around the base of vines can help retain soil moisture and regulate temperature [253]. Additionally, watering early in the morning or late in the evening can minimize evaporation losses [252]. Adjusting watering schedules to account for rainfall is also essential in regions with variable precipitation [252].

Once the vines are established and their root systems extend deeper into the soil, the frequency of watering can be reduced. The focus should shift to deep, infrequent watering that encourages roots to grow downward in search of moisture, thereby improving drought resilience and overall vine health [249].

Initial fertilization of grapevines plays a pivotal role in establishing healthy vines, particularly during the early stages of growth. This process is essential for providing the necessary nutrients that support root development, enhance vine growth, and improve overall vigour. Young grapevines have heightened nutrient demands, and proper fertilization can significantly influence their establishment and long-term productivity.

The initial fertilization of grapevines is critical for several reasons. Firstly, phosphorus is vital for root development, which enables young vines to access water and nutrients more effectively from the soil [254]. Nitrogen is another key nutrient that supports vigorous shoot and leaf growth, laying the groundwork for future fruiting [254]. Additionally, balanced nutrient levels are imperative to prevent deficiencies that can stunt growth and affect vine health [255]. Organic fertilizers, such as compost and manure, not only provide essential nutrients but also enhance soil health by improving its structure and microbial activity, which benefits the vines [256].

Before implementing fertilization strategies, soil testing is imperative to ascertain the existing nutrient profile, pH levels, and organic matter content. Soil tests can reveal baseline levels of key nutrients such as nitrogen, phosphorus, and potassium, as well as the soil pH, which is crucial for nutrient availability [257]. Grapevines thrive in slightly acidic to neutral soils (pH 5.5–7.5), and understanding these parameters allows vineyard managers to tailor their fertilization strategies effectively, avoiding unnecessary applications that could lead to nutrient imbalances or environmental harm [258].

Various types of fertilizers can be utilized during the initial application phase. Organic fertilizers, including compost and bone meal, release nutrients slowly, which is beneficial for sustainable vineyard practices [259]. In contrast, synthetic fertilizers provide precise nutrient delivery, with formulations rich in nitrogen, phosphorus, and potassium being commonly used [254]. Additionally, biofertilizers containing beneficial microorganisms, such as mycorrhizal fungi, can enhance nutrient availability and support vine health, particularly in challenging soil conditions [260].

The primary nutrients required during the initial growth phase include nitrogen (N), phosphorus (P), potassium (K), calcium (Ca), and various micronutrients. Nitrogen is essential for shoot and leaf development, while phosphorus is critical for root establishment [261]. Potassium aids in water regulation and stress tolerance, and calcium contributes to cell wall development [254]. Micronutrients, although needed in smaller quantities, play significant roles in overall vine health and development, necessitating careful monitoring based on soil test results [255].

Effective application methods for fertilizers include pre-planting incorporation, where fertilizers are mixed into the soil before planting, and banding, which concentrates nutrients near the young vines' root zones [261]. Starter solutions can provide an immediate nutrient boost at planting, while topdressing involves applying granular fertilizers around the base of each vine post-planting [254]. These methods ensure that nutrients are readily available for uptake by the young vines.

The timing of fertilization is equally important. A phosphorus-rich fertilizer should be applied at planting to stimulate root growth [254]. During the first growing season, controlled nitrogen applications can promote balanced growth without overstimulating the vines [261]. Regular monitoring of vine growth and adjusting fertilization based on visual symptoms or soil tests is necessary to avoid excessive vegetative growth, particularly late in the season [258].

To optimize initial fertilization, vineyard managers should avoid over-fertilization, which can lead to vigorous growth and increased disease susceptibility [255]. Utilizing organic mulches can help retain soil moisture and improve nutrient availability while suppressing weeds [256]. Regular monitoring for nutrient deficiencies and integrating fertilization with irrigation practices can further enhance nutrient uptake efficiency [257]. Compliance with local agricultural regulations is also essential to ensure sustainable practices and minimize environmental impacts [258].

Initial fertilization is merely the foundation of a comprehensive nutrient management plan. As grapevines mature, their nutrient requirements will evolve, necessitating ongoing soil and tissue testing to refine fertilization strategies [261]. By implementing careful planning and execution of initial fertilization, vineyard managers can establish a strong foundation for young grapevines, ultimately supporting their growth and setting the stage for high-quality grape production in the future.

Chapter 5

Sustainable Vineyard Management

Organic and Regenerative Practices

Organic and regenerative practices in vineyard management are increasingly recognized for their potential to create environmentally sustainable systems that promote biodiversity and enhance soil health. These approaches not only support the production of high-quality grapes but also ensure the long-term viability of vineyards by minimizing synthetic inputs and emphasizing natural processes. The integration of organic and regenerative practices aims to harmonize agricultural activities with the ecosystem, thereby fostering a more resilient agricultural landscape.

Organic viticulture is characterized by the elimination of synthetic chemicals, relying instead on natural methods for pest, disease, and soil fertility management. A fundamental aspect of organic practices is soil management, which prioritizes the maintenance and improvement of soil health through natural amendments. For instance, compost, manure, and green manures are utilized to enrich the soil with organic matter and nutrients, enhancing its fertility and structure [262, 263]. Cover crops, particularly legumes, play a role by fixing nitrogen, reducing erosion, and supporting beneficial soil organisms [264, 265].

Pest and disease control in organic vineyards emphasizes prevention and the use of natural controls. Beneficial insects, such as ladybugs and predatory wasps, are introduced to manage pest populations, while organic-approved substances like neem oil, sulphur, and copper-based sprays are employed to combat fungal diseases [266, 267]. The maintenance of biodiversity within the vineyard ecosystem creates habitats for these natural predators, thereby reducing the likelihood of pest outbreaks [266, 267].

Weed management in organic vineyards eschews chemical herbicides in favour of mulching, mowing, and manual or mechanical weed control. Organic mulches not only suppress weeds but also retain moisture and improve soil health [264, 265]. Nutrient management is achieved through natural sources, including compost and bone meal, with soil testing ensuring balanced nutrient application to avoid overuse that could harm the vines or the environment [263, 265].

Regenerative viticulture extends beyond organic principles by focusing on restoring and enhancing the vineyard ecosystem. This approach emphasizes practices that increase soil organic matter, sequester carbon, and improve biodiversity. For example, regenerative vineyards aim to boost soil carbon levels through practices such as compost application, cover cropping, and reduced tillage, all of which contribute to enhanced soil fertility and water retention [262, 268].

Cover cropping is integral to regenerative farming, as it reduces erosion, fixes nitrogen, and improves water infiltration. Diverse cover crop mixes not only attract pollinators and beneficial insects but also enhance soil structure [264, 268]. Additionally, integrating trees, shrubs, and perennial plants into vineyard landscapes fosters a more diverse ecosystem, providing windbreaks, attracting wildlife, and enhancing microclimates that support vine health and resilience [268, 269].

The incorporation of livestock, such as sheep or chickens, into vineyard management offers natural weed control, fertilization, and pest management. Managed grazing cycles ensure that livestock benefits the vineyard without damaging the vines or compacting the soil [267, 268]. Furthermore, minimizing tillage preserves soil structure and microbial communities, with no-till or minimal-till practices helping to maintain soil health and prevent erosion [264, 270].

Both organic and regenerative practices emphasize efficient water use, employing techniques such as drip irrigation to minimize water waste by delivering precise amounts directly to vine roots. Mulching is also used to reduce evaporation and improve water retention, while swales and contour planting capture rainwater, reducing runoff and increasing infiltration [264].

Creating a biodiverse vineyard environment is a cornerstone of both organic and regenerative viticulture. By encouraging a variety of plant and animal species, growers can establish a balanced ecosystem that supports vine health and reduces the need for external inputs. Habitat creation through wildflower strips, hedgerows, and insect hotels provides homes for pollinators and beneficial insects [266, 271]. Additionally, introducing natural predators, such as birds and mammals, helps control pest populations [266, 267].

Regenerative vineyards contribute to climate change mitigation through carbon sequestration and increased resilience to extreme weather. By building soil organic matter, these vineyards can store atmospheric carbon, reduce greenhouse gas emissions, and improve their capacity to withstand droughts, floods, and temperature extremes [267, 268].

The economic implications of organic and regenerative practices are significant. By reducing reliance on expensive synthetic inputs, vineyard management becomes more cost-effective in the long term. Furthermore, organic and regenerative wines appeal to environmentally conscious consumers, creating opportunities for premium pricing [268, 269]. Community engagement is also fostered through regenerative practices, strengthening relationships and promoting social sustainability [268, 269].

Despite the numerous benefits, organic and regenerative practices present challenges, such as pest and disease management, which require careful monitoring and integrated pest management strategies [267]. Additionally, these practices can be labour-intensive, potentially increasing short-term costs. However, investing in efficient tools and training can help mitigate these challenges [268, 269]. The transition to organic or regenerative viticulture may take several years, necessitating patience and long-term commitment [267, 268].

Composting and Mulching

Compost is a valuable resource in sustainable and organic viticulture, offering benefits such as improved soil health, enhanced nutrient availability, and increased water retention. True compost is a product of a controlled microbiological degradation process, resulting in a friable, dark brown material with a distinct earthy aroma. This process distinguishes compost from simple organic matter left to decay without proper management.

Composting is a deliberate and managed activity requiring uniform mixing of ingredients, maintenance of moisture content between 50% and 60%, adequate oxygen levels, and temperatures ranging from 50°C to 70°C at the pile's core. These conditions support optimal microbial activity, ensuring effective decomposition and the elimination of pathogens and weed seeds. The process can be executed using a variety of methods, from simple windrows turned mechanically to more advanced in-vessel systems. Regardless of the system, consistent management of the composting process is key to producing high-quality compost.

Effective composting relies on balancing carbon-rich and nitrogen-rich materials to achieve a carbon-to-nitrogen ratio between 20:1 and 40:1. Common carbon sources include straw, green waste (tree and garden prunings), paper, and wood, while nitrogen sources include animal manure, food waste, and vegetable scraps. Organic vineyard managers often utilize locally available inputs such as grape marc (a byproduct of winemaking), poultry manure, cattle feedlot manure, sawdust, and other agricultural residues. Additional amendments like rock phosphate and brown coal dust can be incorporated to supply phosphorus and carbon.

Trace elements such as copper, zinc, and manganese, often in sulphate forms, are added based on soil test results to address specific deficiencies. In some cases, lime is mixed with compost to correct soil acidity, saving time and costs during application.

Enhancing Biological Activity

Promoting biological activity in compost can amplify its benefits for the soil. Some growers incorporate microbial inoculants, fish emulsions, and other organic additives that support beneficial bacteria and fungi. These biological enhancements enrich the compost, improving its ability to foster soil health and plant growth.

Figure 25: Biodynamic composting, Granton Vineyard. Stefano Lubiana, CC BY 2.0, via Flickr.

Many vineyard managers adopt a holistic approach to composting, utilizing on-farm resources to close nutrient cycles. For instance, cover crops or other organic materials grown on the property can serve as compost inputs. Livestock integrated into the vineyard ecosystem provide manure that enriches compost, aligning with regenerative and organic certification principles.

Compost can be applied directly to the soil or used as mulch to cover the soil surface. Fine-textured compost is ideal for incorporation into the soil, improving structure, fertility, and microbial activity. Coarser grades of compost or partially decomposed green waste are better suited for surface application, where they act as a mulch to suppress weeds, retain moisture, and regulate soil temperature.

Figure 26: Straw mulch in vineyards in Celler del Roure winery. Artemi Cerdà, Soil Erosion and Degradation Research Team. Universitat de València, Valencia, Spain, CC BY 3.0, via Imaggeo.

To maximize its effectiveness, compost should be tested for nutrient content and composition. A soil analysis before application ensures compatibility with vineyard needs, avoiding potential imbalances like excess potassium from materials such as grape marc.

While some growers produce their compost using local materials, others source commercially available compost. Commercial compost products should meet established standards to ensure they are free of contaminants and safe for plant growth. For example, in Australia, the voluntary standard AS 4454-1997 outlines criteria for compost quality. Vineyard managers should request documentation from suppliers detailing the compost's ingredients, nutrient content, and compliance with organic certification standards if applicable.

Compost offers numerous advantages beyond nutrient supplementation. It improves soil structure, enhances water infiltration and retention, supports microbial diversity, and reduces soil compaction. By increasing organic matter levels, compost also boosts the soil's ability to sequester carbon, contributing to climate change mitigation. While compost is not a complete substitute for fertilizers, it complements them by enhancing soil health and providing a slow-release source of nutrients.

Compost Application in Vineyards

Applying compost in vineyards requires careful planning and analysis to ensure that the nutrients provided align with the vineyard's needs without causing over-fertilization. The application rate depends on the compost's nutrient content, the vineyard's soil and plant nutrient status, and the desired outcome for vine growth and grape quality. This process emphasizes the importance of regular soil and petiole nutrient analyses to guide compost application decisions [272].

Compost application rates are primarily determined by the nitrogen content of the compost, as nitrogen has a significant impact on vine growth. While nitrogen is essential for healthy vine development, excessive levels can stimulate excessive vegetative growth, leading to shading, reduced fruit quality, and increased disease pressure. If soil and plant analyses show adequate or high nitrogen levels, additional compost may not be necessary and could exacerbate these issues. Conversely, if nitrogen levels are low and vine growth is suboptimal, a measured compost application can be beneficial, providing a gradual release of nutrients over time [272].

Typically, about 30% of the total nitrogen in compost becomes available for vine uptake over several years—15% in the first year and the remaining 15% over the next three to four years. For instance, applying 10 tons of compost containing 20 lbs of nitrogen per ton would provide approximately 30 lbs of nitrogen per acre in the first year, with another 30 lbs available gradually over the subsequent years. Despite these incremental nutrient releases, growers should note that 10 tons of compost per acre often appear minimal on the soil surface, underscoring the importance of balancing visible coverage with nutrient needs [272].

Growers are advised to carefully observe the effects of initial compost applications on vine vigour and grape quality before determining future application rates. Over-application may lead to over-fertilization, which can negatively impact grape quality and increase production costs [272].

The method of compost application depends on the vineyard's size, layout, and the intended coverage area. Small vineyards or specific sections can be fertilized manually, ensuring precise placement of compost around the vines. For larger vineyards, machinery such as broadcast manure spreaders can evenly distribute compost across the vineyard floor. These machines are particularly effective for pre-planting applications or when composting the entire vineyard area [272].

Alternatively, mulch spreading machines can facilitate band applications under the vine rows, targeting the root zone while leaving the inter-row areas untreated. This method is suitable for growers focusing on vine-specific fertilization rather than widespread nutrient distribution [272].

Timing is an important aspect of compost application. The ideal periods for application are in the fall, immediately after harvest, or early spring, just before bud break. These timings ensure that nutrients are available when the vines need them most—during bud break and the early growing season—while avoiding excess vine vigour later in the season. Mid-season applications are generally discouraged, as they can promote excessive vegetative growth during a period when vines should be focusing on fruit development and acclimation for winter [272].

No-Till Farming Techniques

No-till farming techniques in viniculture represent a transformative approach to vineyard management that prioritizes soil health, biodiversity, and sustainability. This method minimizes soil disturbance, which is crucial for maintaining soil structure and fostering a robust ecosystem that supports grape production. By avoiding traditional tillage practices, no-till farming enhances the natural integrity of the soil, allowing for improved nutrient cycling and plant health, which are essential for sustainable viticulture [262, 264, 273].

The core principles of no-till farming in vineyards include minimizing soil disturbance, maintaining soil cover, and enhancing biodiversity. Minimizing soil disturbance is vital as it preserves the soil's structure and prevents compaction, which can inhibit root growth and water infiltration [274]. Maintaining a protective layer of organic matter, such as cover crops or mulch, is essential for preventing erosion, retaining moisture, and regulating soil temperature [264, 275]. Furthermore, enhancing biodiversity through the encouragement of various plant species and beneficial organisms reduces reliance on synthetic inputs, thus promoting a more resilient vineyard ecosystem [276, 277].

The benefits of no-till farming techniques are manifold. Improved soil structure and health are achieved as soil aggregates remain intact, allowing for better root penetration and water infiltration [264, 273, 278]. The organic layer on the soil surface significantly enhances water retention, which is particularly beneficial during dry periods [264, 279]. Additionally, no-till practices contribute to a reduction in soil erosion, especially in sloped vineyards, where ground cover from mulch or cover crops is crucial [280]. The practice also promotes carbon sequestration, which is vital for mitigating climate change impacts [273]. Moreover, a no-till environment supports beneficial organisms such as earthworms and mycorrhizal fungi, which improve nutrient availability and suppress pathogens [281]. Lastly, the reduction in mechanical tillage leads to lower costs and energy use, making no-till farming an economically viable option for vineyard management [273, 277].

Several key techniques are employed in no-till viniculture, including the use of cover crops, mulching, and integrated weed control. Cover crops, such as clover and rye, are planted between vineyard rows to protect the soil, fix nitrogen, and suppress weeds [262, 282]. Mulching with organic materials like straw or wood chips conserves moisture and enhances soil organic content [264, 275]. Undervine management techniques, utilizing specialized tools to manage vegetation without disturbing the soil, are also critical [283]. Compost application is another technique that adheres to no-till principles by enriching the soil while preserving its structure [273, 284]. Integrated weed control strategies, which rely on natural suppressants rather than synthetic herbicides, are essential for maintaining a healthy vineyard ecosystem [285].

Despite the numerous benefits of no-till farming, challenges remain. Weed pressure can be a significant concern, as effective control relies on careful planning and the use of cover crops or mulches [283]. Pest management can also be complicated by the organic debris layer, necessitating integrated pest management strategies to mitigate potential issues [276, 277]. Additionally, nutrient cycling may be slower due to the decomposition of organic materials, requiring adjustments in fertilization practices to ensure adequate nutrient availability [273, 286].

The adoption of no-till farming techniques in vineyards is gaining traction globally, from European wine regions to vineyards in the Americas and Australia. This method aligns with sustainability goals and organic certification requirements, addressing environmental challenges such as soil erosion, drought, and biodiversity loss [273, 279]. With proper implementation, no-till farming can enhance grape production quality while ensuring long-term ecological balance and soil vitality [262, 284].

Integrated Pest and Disease Management (IPDM)

Pests and diseases have a profound impact on grapevines, influencing their growth, yield, quality, and longevity. The severity of these effects depends on the specific pest or disease, environmental factors, and the management practices employed by vineyard operators.

Pests can severely affect vine growth and vitality. Sap-feeding pests such as mealybugs, aphids, and scale insects drain the vines of essential nutrients by feeding on their sap, leading to reduced vigour and productivity. Root-feeding pests like grape phylloxera and nematodes target the roots, disrupting the plant's ability to absorb water and nutrients, which stunts growth and can cause vine death. Leaf-feeding pests, including caterpillars, thrips, and Japanese beetles, damage leaves and reduce the vine's capacity for photosynthesis, leading to slower growth and weakened overall health. Diseases further compound these challenges. Fungal infections like powdery mildew, downy mildew, and black rot attack the vine's foliage, shoots, and berries, impairing photosynthesis and weakening the plant. Trunk diseases such as Esca and Eutypa dieback infect the woody parts of the vine, disrupting the flow of water and nutrients and causing vine dieback, which significantly hampers productivity.

Pests and diseases also lead to reduced yields and diminished fruit quality. Pests such as grape berry moths and spider mites directly damage grape clusters, reducing the number of viable berries and lowering overall yields. Diseases like botrytis bunch rot and black rot can cause berries to shrivel and decay, rendering them unmarketable or unusable. The quality of the fruit is often compromised, as pests and diseases can leave berries susceptible to secondary infections, reducing their flavour, sugar content, and overall quality. Viral infections such as Grapevine Leafroll Virus disrupt sugar accumulation in berries, leading to uneven ripening and poor-quality grapes. Additionally, diseases like botrytis can cause undesirable off-flavours in wine, such as mouldy or musty notes.

Economically, the impact of pests and diseases on grapevines is significant. Managing these challenges involves regular monitoring, the use of biocontrol agents, and the application of pesticides and fungicides, all of which increase operating costs. Severe infestations or infections can result in vine loss, necessitating replanting and further expenses for vineyard rehabilitation. Additionally, the reduced quality of grapes affects market prices, directly impacting the profitability of the vineyard.

Over the long term, pests and diseases can compromise vineyard health. Viral diseases like Grapevine Leafroll Virus are often spread by vectors such as mealybugs, making containment difficult and potentially reducing the lifespan of the vines. Root-feeding pests and diseases can degrade soil health, diminishing its suitability for future grapevine growth. Excessive pesticide use to manage pests can harm beneficial insects and microorganisms, reducing biodiversity and making the vineyard ecosystem more vulnerable to future outbreaks.

The environmental and social impacts of pests and diseases are also significant. Persistent pest and disease issues often lead to reliance on chemical controls, which can negatively affect surrounding ecosystems and increase the carbon footprint of viticulture. Managing outbreaks requires additional labour for monitoring, pruning, and treatment, placing strain on vineyard operations, particularly in small-scale or family-run vineyards.

Pests and diseases present a constant threat to grapevine health, affecting every aspect of viticulture, from vine growth to grape quality and vineyard profitability. Employing Integrated Pest and Disease Management (IPDM) and sustainable practices is crucial for mitigating these effects. Through early detection, prevention, and a balanced approach that includes biological, cultural, and chemical controls, vineyards can maintain productivity while producing high-quality grapes.

Major Pests Affecting Grapevines

Grape Phylloxera (*Daktulosphaira vitifoliae*): Phylloxera is a small aphid-like insect that feeds on the roots of grapevines, causing galls and disrupting the flow of nutrients and water. Native to North America, it has spread globally, devastating vineyards with susceptible rootstocks. **Control**: Grafting European grapevines (*Vitis vinifera*) onto resistant American rootstocks is the most effective solution.

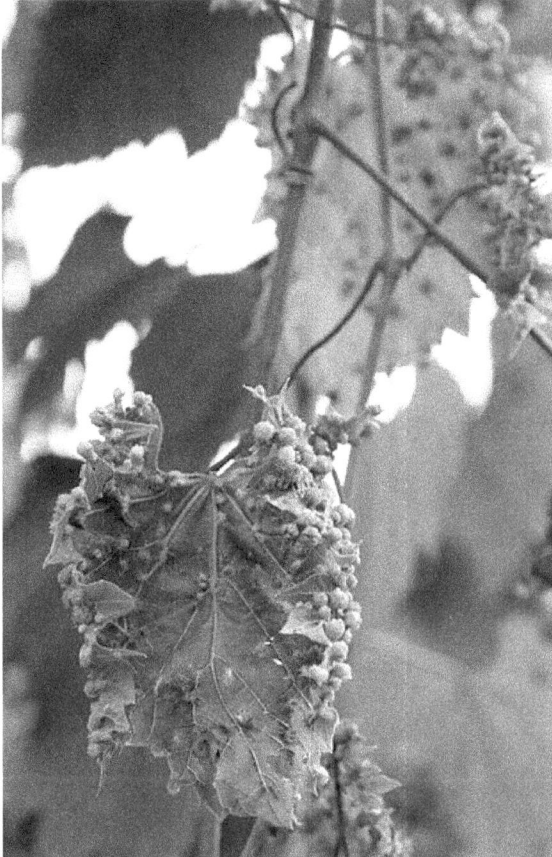

Figure 27: A grape leaf with galls formed during a phylloxera infection. cathdrwg, CC BY 2.0, via Wikimedia Commons.

Mealybugs: Mealybugs infest vines by feeding on plant sap, leading to reduced vigour. They also excrete honeydew, which encourages sooty mould growth and serves as a vector for viral diseases like Grapevine Leafroll Virus. **Control**: Biological control with predators such as lady beetles and parasitoid wasps, along with targeted insecticides, can manage infestations.

Grape Berry Moth (*Lobesia botrana*): The larvae of the grape berry moth feed on grape clusters, causing direct damage and increasing susceptibility to fungal infections like Botrytis. **Control**: Pheromone traps for monitoring and mating disruption, combined with insecticides, are effective in managing populations.

Figure 28: Paralobesia viteana – Grape Berry Moth. Andy Reago & Chrissy McClarren, CC BY 2.0, via Wikimedia Commons.

Spider Mites: Spider mites feed on grapevine leaves, causing stippling, yellowing, and reduced photosynthetic capacity. Severe infestations can weaken the vine and reduce yield. **Control**: Integrated Pest Management (IPM) using predatory mites and miticides.

Thrips: Thrips feed on young shoots, flowers, and fruit, causing scarring and deformities. While minor infestations may not cause significant damage, severe cases can affect crop quality. **Control**: Monitoring and biological controls, along with selective insecticides, can manage populations.

Japanese Beetle (*Popillia japonica*): This invasive pest feeds on grapevine leaves, skeletonizing them and reducing photosynthesis. It is most common in North America. **Control**: Handpicking in small vineyards, pheromone traps, and targeted insecticides can help reduce damage.

Major Diseases Affecting Grapevines

Powdery Mildew (*Erysiphe necator*): Powdery mildew affects leaves, shoots, and fruit, leaving a white, powdery coating. It thrives in warm, humid conditions and can severely impact yield and wine quality. **Control**: Regular applications of sulphur or other fungicides, canopy management to improve airflow, and resistant cultivars.

Downy Mildew (*Plasmopara viticola*): Downy mildew causes yellow, oily spots on leaves, leading to defoliation and reduced photosynthesis. Severe infestations can result in crop loss. **Control**: Copper-based fungicides, good drainage, and canopy management to reduce leaf wetness.

Figure 29: Mildew on 'Valiant' cultivar (Vitis vinifera x Vitis labrusca x Vitis riparia), Sporangia and sporangiophores protruding from the stomas of the grape berries. David S. Jones, Patricia McManus, CC BY-SA 4.0, via Wikimedia Commons.

Botrytis Bunch Rot (Botrytis cinerea): Botrytis affects ripening grape clusters, causing grey mould. It is particularly problematic in humid conditions and can lead to significant crop loss. **Control**: Proper canopy management, fungicides, and the use of resistant grape varieties.

Black Rot (*Guignardia bidwellii*): This fungal disease causes brown leaf spots and black, shriveled berries, reducing yield and quality. It thrives in warm, wet conditions. **Control**: Pruning infected material, fungicides, and maintaining good vineyard hygiene.

Esca (Grapevine Trunk Disease): Esca is a complex disease caused by a group of fungi that infect grapevine wood, leading to vine dieback. Symptoms include tiger stripe patterns on leaves and berry shrivel. **Control**: Removing infected vines, avoiding pruning wounds during wet conditions, and applying protective treatments to pruning cuts.

Figure 30: Esca - Infestation of a grapevine. Bauer Karl, CC BY 3.0 AT, via Wikimedia Commons.

Grapevine Leafroll Virus: This viral disease slows vine growth, reduces berry ripening, and alters grape composition, affecting wine quality. It is spread by mealybugs and scale insects. **Control**: Using certified virus-free planting material, controlling vector populations, and removing infected vines.

Pierce's Disease (*Xylella fastidiosa*): Pierce's Disease is caused by a bacterium spread by sharpshooters, particularly the glassy-winged sharpshooter. It blocks water transport within the vine, leading to leaf scorching and vine death. **Control**: Removing infected vines, managing sharpshooter populations, and planting tolerant rootstocks in affected regions.

Crown Gall (*Agrobacterium tumefaciens*): Crown gall causes tumour-like growths at the base of grapevines, disrupting nutrient flow and reducing vigour. It is more severe in vines stressed by cold damage. **Control**: Planting disease-free material, avoiding mechanical injuries, and removing infected vines.

Integrated Pest and Disease Management (IPDM)

The most effective way to manage grapevine pests and diseases is through IPDM, which combines cultural, biological, and chemical control measures. Strategies include:

- Regular monitoring for early detection of pests and diseases.

- Encouraging biodiversity in and around the vineyard to support natural predators.

- Using certified planting material to prevent disease introduction.

- Applying targeted treatments based on economic thresholds to minimize environmental impact.

- Practicing good vineyard hygiene by removing infected plant material and sanitizing tools.

By adopting region-specific IPDM strategies and sustainable practices, grape growers worldwide can effectively manage pests and diseases, ensuring healthy vineyards and high-quality grape production.

Integrated Pest and Disease Management (IPDM) in vineyards is an essential strategy for maintaining ecological balance while ensuring sustainable viticulture. This approach integrates various tactics aimed at controlling pests and diseases, thereby minimizing their impact on grape production and the environment. The core tactics of IPDM include biological control, cultural practices, chemical control, economic thresholds, and mass trapping devices, each contributing to the overarching goal of sustainable vineyard management.

Biological control is a fundamental aspect of IPDM, utilizing natural predators and parasitoids to manage pest populations. For instance, the introduction of predatory mites and parasitic wasps has been shown to effectively control pests such as spider mites and grapevine moth larvae [287, 288]. Research indicates that enhancing the habitat for these beneficial organisms can significantly improve pest control outcomes in vineyards [289, 290]. Moreover, studies have demonstrated that vineyards with diverse habitats support higher populations of natural enemies, which can lead to reduced reliance on chemical pesticides [291, 292].

Cultural practices also play a crucial role in IPDM. These practices involve the selection of pest-resistant grape varieties, proper sanitation, and vineyard management techniques such as pruning and canopy management to improve air circulation and reduce disease incidence [285, 293]. For example, maintaining ground cover and managing inter-row vegetation can enhance biodiversity, which in turn supports beneficial arthropods that contribute to pest control [294, 295]. Additionally, the use of green manures has been linked to improved soil health and increased populations of natural pest controllers [296].

Chemical control, while considered a last resort, is still a component of IPDM. The judicious use of selective pesticides, such as sulphur-based fungicides for powdery mildew, is recommended to minimize adverse effects on non-target species [289, 297]. The timing and application methods of these chemicals are critical to reducing environmental impacts and preventing the development of pest resistance [298]. Regular monitoring of pest populations allows vineyard managers to apply chemical controls only when necessary, aligning with the economic threshold concept that defines the pest density at which control measures become economically justified [299, 300].

Mass trapping devices, particularly pheromone traps, are effective in managing specific pests like the European grapevine moth. These traps disrupt mating cycles and capture significant numbers of pests, thereby reducing their reproductive potential without resorting to chemical interventions [301]. This method exemplifies the integration of monitoring and control strategies within the IPDM framework.

The IPDM framework emphasizes a triad of prevention, monitoring, and control. Prevention strategies focus on establishing healthy vineyard practices, such as selecting appropriate sites and maintaining soil health [289, 299]. Monitoring is vital for the success of IPDM, utilizing tools like visual inspections and pheromone lures to assess pest populations and predict disease outbreaks based on environmental conditions [299]. When pest populations exceed economic thresholds, control measures are implemented, prioritizing biological and cultural practices before resorting to chemical controls [289, 302].

The implementation of IPDM in vineyards offers numerous benefits. Environmentally, it reduces pesticide use, thereby preserving biodiversity and protecting non-target species, including pollinators [289, 290]. Economically, it minimizes input costs by ensuring that interventions are only applied when necessary, thus enhancing the sustainability of vineyard operations [285, 303]. Furthermore, IPDM contributes to resistance management by preventing pests from developing resistance to chemical controls through reduced reliance on pesticides [289, 297].

Globally, IPDM is adapted to various vineyard conditions, reflecting local ecological and regulatory contexts. In Europe, stringent pesticide regulations have made IPDM a critical tool for compliance and sustainability [289, 302]. In North America, IPDM supports organic and sustainable certifications, which are increasingly sought after by consumers [289]. In tropical regions, where pest pressures are high, IPDM emphasizes biological controls and cultural practices to manage these challenges effectively [289, 303].

Biological and Biorational Pest Control in Vineyards

Effective pest control in vineyards is essential for maintaining vine health, ensuring grape quality, and achieving sustainable viticulture. Biological and biorational pest control methods have become increasingly popular worldwide as eco-friendly alternatives to conventional synthetic pesticides. These methods emphasize the use of natural agents and environmentally benign substances to manage pests while preserving the ecological balance of the vineyard.

Biological pest control leverages natural enemies of pests to reduce their populations in vineyards. This method involves introducing or encouraging beneficial organisms such as predatory mites, pirate bugs, soil-dwelling mites, and parasitic insects. These organisms prey on or parasitize harmful pests, reducing their impact on grapevines.

Predatory mites, for instance, feed on spider mites, a common vineyard pest that can damage leaves and reduce photosynthesis. Pirate bugs are effective against thrips and small insects, while parasitic wasps target the larvae of moths and other pests. Soil-dwelling mites contribute to controlling pests like nematodes, which can attack vine roots.

Biological control agents do not entirely eliminate pest populations but aim to maintain them at levels that do not threaten vine health or productivity. This approach aligns with sustainable and organic viticulture, reducing reliance on chemical interventions and preserving biodiversity within the vineyard ecosystem.

Biorational pest control involves using substances that are less toxic to the environment, target-specific, and degrade rapidly after application. Unlike conventional pesticides, which often affect a broad range of organisms, biorationals are designed to minimize harm to non-target species, including beneficial insects and microorganisms.

Biorationals are categorized based on their shared characteristics rather than their chemical composition. They often have highly specific modes of action, targeting particular pests or disrupting specific biological processes. For example, some biorational pesticides interfere with pest reproduction, feeding, or development rather than causing direct mortality.

This method is particularly valuable in vineyards where maintaining the delicate balance of pest and predator populations is critical. By using biorationals, vineyard managers can effectively control pests while safeguarding beneficial organisms that contribute to the vineyard's ecological health.

Biopesticides represent a subset of biorationals and are derived from natural materials such as animals, plants, bacteria, and minerals. These products are often used in vineyards worldwide to manage pests in an environmentally friendly manner. Biopesticides are categorized into three main types:

1. **Biochemical Pesticides:** These naturally occurring substances control pests through non-toxic mechanisms. For example, sex pheromones can be used to disrupt mating cycles, reducing pest reproduction. Similarly, scented plant extracts can attract pests to traps, lowering their population without harming the environment.

2. **Microbial Pesticides:** These consist of microorganisms like bacteria, fungi, viruses, or protozoa that act as active ingredients. For instance, *Bacillus thuringiensis* (Bt) is a bacterium used to target specific pest larvae while being safe for other organisms. Fungal pathogens can also infect and control insect pests, contributing to the vineyard's pest management strategy.

3. **Plant-Incorporated Protectants (PIPs):** These are pesticidal substances produced by crops that have been genetically engineered to contain pest-resistant traits. For example, certain grape varieties may be developed to produce proteins toxic to specific pests, reducing the need for external pesticide applications.

Many biorationals are approved for use in organic viticulture, making them a natural fit for managing pests in vineyards committed to organic farming practices. However, not all formulations meet the stringent requirements of the National Organic Program (NOP) or equivalent global standards. For example, phosphorus acids and genetically engineered PIPs are typically not allowed under organic certification.

Despite these limitations, approved biorationals provide organic growers with effective tools to manage pests while maintaining compliance with certification requirements. Products such as neem oil, insecticidal soaps, and microbial biopesticides are widely used in organic vineyards, offering targeted pest control with minimal environmental impact.

Biological and biorational pest control methods are applicable across diverse viticultural regions worldwide. In Europe, where stringent regulations limit synthetic pesticide use, these methods play a crucial role in sustainable vineyard management. In North and South America, the increasing consumer demand for organic and sustainably produced wines has driven the adoption of these eco-friendly practices. In regions like New Zealand and South Africa, where biodiversity and environmental preservation are key priorities, biological and biorational approaches align with national and regional sustainability goals.

By integrating these methods into vineyard management practices, viticulturists can achieve effective pest control, enhance ecological resilience, and produce high-quality grapes while minimizing environmental impact. This approach reflects the growing global commitment to sustainable and regenerative agriculture.

Chemical Pest Control

Chemical pest control is an essential tool in viticulture and agriculture, particularly when other methods in an Integrated Pest Management (IPM) strategy fail to keep pest populations below economic thresholds. The use of synthetic or organic pesticides requires careful consideration of environmental, economic, and health impacts to ensure sustainable farming practices while minimizing negative consequences.

Conventional pesticides are synthesized chemical compounds designed to control a wide array of agricultural pests, including insects, weeds, and fungal pathogens. These pesticides fall into several main classes, each with distinct properties and modes of action. Organochlorines, known for their long-lasting effects, are largely banned or restricted in many countries due to their environmental persistence and toxicity. Organophosphates are effective against a broad range of pests but pose significant risks to human and animal health because of their neurotoxic effects. Carbamates, which function similarly to organophosphates, offer a shorter environmental persistence, making them a preferable choice in some cases. Pyrethroids, synthetic derivatives of naturally occurring pyrethrins, provide an effective and less toxic alternative to older chemical classes while maintaining high efficacy. These synthetic compounds are primarily used in conventional farming, forming the backbone of pest control strategies in vineyards where high-value crops face significant risks. Their application is justified only when pest populations surpass economic injury levels, threatening significant yield losses.

Chemical control is often considered a last resort in the IPM framework. IPM prioritizes cultural, biological, and mechanical controls to manage pest populations sustainably. However, when these methods prove insufficient, pesticides become a necessary intervention to prevent economic losses. To maximize efficacy and minimize adverse impacts, pesticides should be applied selectively to target specific pests without harming beneficial organisms such as pollinators and natural predators. Precise timing of applications is critical for optimal control while minimizing unnecessary chemical exposure to the environment. Adhering to recommended dosages and proper application techniques helps prevent pesticide resistance and environmental contamination.

In organic farming systems, chemical pest control is regulated by strict guidelines that prioritize natural and minimally processed substances. The National Organic Program (NOP), managed by the USDA, outlines specific regulations for using these chemicals. Organic-certified growers may use substances derived from natural sources or certain approved synthetic chemicals listed in the National List of Allowed and Prohibited Substances. Common organic-approved substances include sulphur, a natural fungicide effective against powdery mildew and other fungal diseases, and copper compounds, which are widely used for managing fungal infections but require careful management to prevent soil accumulation. Neem oil, derived from the neem tree, acts as a plant-based insecticide and miticide, while Bacillus thuringiensis (Bt), a microbial insecticide, targets caterpillars and other larvae without harming non-target species. Organic farmers must demonstrate that chemical interventions are part of a broader strategy incorporating crop rotation, biological controls, and cultural practices, using these substances only when all other preventive measures fail.

While chemical pest control is an effective tool, its usage entails responsibilities to mitigate risks to the environment and human health. Conventional pesticides, despite their efficacy, can have long-term consequences such as soil degradation, water contamination, and disruption of beneficial ecosystems. Similarly, organic-approved substances, though derived from natural sources, must be used carefully to prevent overapplication and unintended environmental impacts. Growers worldwide must navigate varying regulatory landscapes, public perceptions, and specific pest pressures unique to their regions. Sustainable pest control requires integrating chemical methods into holistic management plans that prioritize long-term vineyard health, economic viability, and environmental stewardship.

Fungal Disease Prevention

Fungal disease prevention in vineyards is critical for maintaining healthy vines and ensuring high-quality grape production. Fungal pathogens, such as powdery mildew, downy mildew, botrytis bunch rot, black rot, and Eutypa dieback, can inflict significant damage on grapevines, leading to reduced yields, compromised grape quality, and even vine death. These pathogens thrive in warm, humid, or wet conditions, spreading through spores carried by wind, water, or vineyard equipment. Once established, they can infect various parts of the vine, including leaves, stems, and grape clusters, which diminishes photosynthesis, causes fruit decay, and weakens the vines [304, 305].

Effective prevention of these diseases involves a multifaceted approach that includes cultural, biological, and chemical strategies tailored to the specific conditions of each vineyard. Cultural practices are foundational in creating an environment less conducive to fungal growth. For instance, proper canopy management through pruning, training, and shoot positioning enhances air circulation and light penetration, which helps to reduce humidity and promote quicker drying of surfaces after rainfall or irrigation [306, 307]. Additionally, sanitation practices, such as the removal of infected plant material, significantly reduce the reservoir of fungal spores, particularly for diseases like black rot and botrytis, which can overwinter in debris [308, 309]. Site selection and design also play a crucial role; vineyards planted on well-drained soils with good sun exposure can minimize moisture levels that favour fungal growth [310].

Biological controls represent another effective strategy for managing fungal diseases. The application of beneficial microorganisms, such as Trichoderma species, can outcompete harmful fungal pathogens or protect vine wounds from infection [311, 312]. Organic sprays, including compost teas and plant extracts, can bolster the plant's natural defences against fungal pathogens [313]. These biological agents are often integrated into a broader disease management program to enhance overall vineyard health.

When cultural and biological measures are insufficient, chemical controls, particularly fungicides, become necessary. Preventative fungicides, such as sulphur and copper-based products, are commonly employed in both conventional and organic vineyards to avert fungal infections. These fungicides must be applied before infection occurs and reapplied based on weather conditions and disease pressure [170, 314]. Systemic fungicides, which penetrate plant tissues for longer-lasting protection, are also utilized in conventional viticulture, often rotated with other products to prevent resistance development [315, 316]. The timing and application of these fungicides are critical, with applications made during key growth periods, such as bud break and flowering, to maximize their effectiveness [317].

Regular monitoring of weather conditions, vine health, and fungal symptoms is essential for effective prevention. Many vineyards utilize weather stations or disease forecasting models to predict outbreaks based on environmental factors such as temperature and humidity. This proactive approach allows for optimized timing of interventions, reducing unnecessary fungicide applications and enhancing sustainability.

Integrated Disease Management (IDM) combines these various strategies into a holistic approach. By minimizing reliance on any single control method, IDM reduces the risk of resistance development and environmental impact while maintaining vine health and productivity. This comprehensive strategy emphasizes the importance of cultural and biological controls, judicious chemical interventions, and continuous monitoring to adapt to changing conditions in the vineyard.

Water Conservation and Irrigation Strategies

Water conservation and irrigation strategies in viniculture are vital for sustainable vineyard management, especially in regions with water scarcity or unpredictable climate patterns. Effective water management not only ensures optimal vine health and enhances grape quality but also reduces the environmental impact of viticulture practices. Balancing water usage with the needs of the vineyard is crucial for maintaining long-term sustainability.

Water plays a central role in vine physiology, influencing growth, fruit development, and overall health. Insufficient water can cause vine stress, which negatively impacts grape yield and quality. On the other hand, over-irrigation can lead to excessive vegetative growth, making the vines more susceptible to diseases and nutrient leaching. Achieving efficient water management is therefore essential for balancing vine growth with grape composition, ensuring both productivity and quality.

Soil moisture management is a key technique for conserving water in vineyards. Practices such as mulching with organic materials like straw, compost, or wood chips help retain soil moisture, reduce evaporation, and moderate soil temperatures. Planting cover crops between vine rows improves water infiltration and reduces surface runoff, while carefully preparing soil structure enhances its ability to retain water. Additionally, canopy management techniques, such as pruning, leaf removal, and shoot thinning, optimize sunlight exposure and airflow, reducing unnecessary transpiration and overall water loss. Smaller, well-maintained canopies require less water and contribute to balanced grape ripening.

Irrigation scheduling is another critical component of water conservation. Tools like soil moisture sensors, tensiometers, and evapotranspiration data allow precise scheduling of irrigation events, ensuring water is applied only when vines need it. Vineyard design also plays a significant role, with practices such as planting on contours or using terracing to minimize water runoff and enhance moisture retention. Selecting drought-tolerant rootstocks and grape varieties adapted to local climatic conditions further supports water conservation efforts.

Efficient irrigation systems are integral to sustainable water management in vineyards. Drip irrigation is widely recognized as the most water-efficient method, delivering water directly to the root zone through emitters. This minimizes evaporation and runoff while allowing precise control over water application. Subsurface drip irrigation systems place drip lines below the soil surface, further reducing evaporation and delivering water directly to the root zone, which is particularly effective in arid regions. Regulated deficit irrigation (RDI) intentionally supplies less water than vines' full requirements during specific growth stages, promoting deeper root growth and enhancing grape quality by concentrating flavours and sugars. Partial rootzone drying (PRD) alternates water supply between halves of a vine's root system, reducing overall water usage while maintaining productivity and promoting flavour compound development.

Water recycling and rainwater harvesting provide sustainable options for irrigation. Vineyards can install systems to collect and store rainwater, reducing dependency on external water sources. Recycling wastewater from winemaking processes, once treated to remove contaminants, is another viable approach. These methods allow vineyards to utilize water resources efficiently, especially during dry periods.

Technological advancements have further enhanced water management in viniculture. Remote sensing, drones, and satellite imagery offer real-time data on vine water stress, soil moisture, and canopy conditions, allowing for informed decisions. Automated irrigation systems controlled through smartphone applications provide precise water delivery based on sensor inputs, improving efficiency and reducing waste.

With climate change intensifying water scarcity in many wine-growing regions, vineyards must adopt adaptive strategies. These include planting drought-resistant grape varieties, adjusting planting schedules, and investing in infrastructure to capture and store water. Innovative techniques, such as using hydrogels to retain soil moisture, further support these adaptation efforts. Combining traditional practices with modern technology enables vineyards to address water-related challenges effectively while maintaining productivity and sustainability.

Managing Grapevines with Less Water

Managing grapevines with reduced water availability due to dry seasonal conditions presents a complex challenge for vineyard operators. Irrigators must evaluate a range of strategies to mitigate the impact of water scarcity while balancing immediate and long-term production goals. Decisions may include purchasing additional water in a competitive market, prioritizing water allocation to specific varieties or high-value vineyard patches, and reducing water supply to lower-priority areas, which may lead to decreased yields. In extreme cases, poorly performing patches might be abandoned, or low-priority plantings removed altogether, with redevelopment plans accelerated.

These decisions are crucial not only for surviving drought conditions but also for ensuring vineyard recovery and productivity in subsequent seasons.

Grapevines are uniquely resilient and can recover from extended periods of low water availability more effectively than many other crops, making them a preferred candidate for irrigation prioritization in mixed horticultural enterprises. However, water management strategies must account for the specific water requirements of grapevines across their five key development stages, as each stage has different implications for vine growth, yield, and long-term health.

From pre-budburst to flowering, grapevines use relatively little water. This stage is critical for determining potential yield, and ensuring adequate moisture at flowering is essential. A full soil-moisture profile at budburst fosters early canopy development, which protects developing bunches later in the season. During flowering and fruit set, vines are highly sensitive to water stress, as this is a period of rapid shoot growth. Insufficient water at this stage can significantly reduce yield and impede vine development. If spring rainfall is adequate and soil moisture is near field capacity at budburst, irrigation may be delayed until mid-October.

The period from fruit set to veraison accounts for approximately a third or more of total seasonal water use, particularly in winegrape cultivation. Water deficit strategies, such as regulated deficit irrigation (RDI), are commonly employed during this stage to control vine vigour and improve grape quality, especially in varieties like Shiraz that produce large berries. However, these strategies may lead to reduced berry size, which is often acceptable for winegrape production. From veraison to harvest, another substantial portion of seasonal water is used. Water stress during this period can decrease sugar accumulation, adversely affecting grape ripening and quality. After harvest, during the leaf-fall period, water use is relatively low, but excessively dry soils can hinder carbohydrate accumulation and nutrient uptake, impacting vine health and productivity in the following season.

The impact of reduced water availability varies significantly across grape types. Winegrapes, having been the focus of extensive research, offer the most flexibility in deficit irrigation practices. Water deficits during the fruit set to veraison stage are generally considered suitable for winegrape varieties, especially those with a propensity for large berries. By contrast, table grapes require significantly more water to maintain their marketable size and quality. Any reductions in water supply for table grapes typically result in smaller berries and unmarketable bunches, leaving growers with few options other than prioritizing patches or purchasing additional water.

Dried vine fruit, such as raisins, has different water requirements based on production methods. Modern management practices like swing-arm trellises demand higher water inputs due to their larger canopies compared to traditional methods. While deficit irrigation can be applied during the fruit set to veraison stage with less dramatic effects than those seen in table grapes, it may still reduce berry size. Following the cutting of the fruiting canes, water requirements decrease, offering an opportunity to conserve water, though the reduction in water demand may not be as significant as previously assumed.

In all cases, efficient irrigation practices, precise scheduling, and adaptive strategies are critical to managing vineyards under water scarcity. Understanding the specific needs of each grape type and development stage ensures that growers can make informed decisions to optimize resource use while mitigating the impacts of drought conditions.

Regulated Deficit Irrigation (RDI) is an advanced irrigation strategy used in viticulture to enhance the quality of red wine grapes by carefully managing water stress during specific growth phases. The technique involves applying less water than the vines require for unrestricted growth, typically during the period from fruit set to veraison. By doing so, RDI moderates vine canopy growth and reduces berry size. Smaller berries are desirable for winemaking as they have a higher skin-to-juice ratio, enhancing the concentration of phenolics and flavours, which contribute to improved wine quality.

The success of RDI depends on achieving a delicate balance—enough water stress to influence berry size and composition positively, but not so much as to severely impact vine health or long-term productivity. Despite its potential benefits, the adoption of RDI has declined in some regions as the reduction in yields has not been compensated by higher returns for wine grapes. However, when water availability is limited, RDI remains a valuable option, especially during the fruit set to veraison phase, when the vine's response to water deficits can significantly impact grape quality.

Sustained Deficit Irrigation (SDI), in contrast, applies a consistent reduction in water across the entire growing season rather than targeting specific growth phases. This approach delivers a lower volume of water at each irrigation event, creating a sustained water deficit. SDI has the potential to conserve water in regions where availability is limited while maintaining reasonable production levels. However, outcomes from SDI trials have shown variability depending on grapevine varieties and the severity of the water deficit. For example, first-season trials often show minimal yield reduction, as fruit production depends on the previous season's growth under full irrigation. Over successive seasons, however, significant yield declines have been observed in varieties like Cabernet Sauvignon and Shiraz when subjected to 50% SDI.

Research suggests that a deficit irrigation level equal to or greater than 75% of crop evapotranspiration (ETc), whether applied as RDI or SDI, can sustain near-optimal production for certain cultivars like Sunmuscat in the short to medium term. These findings highlight the importance of tailoring irrigation strategies to the specific needs and responses of different grapevine varieties under water-limited conditions.

Rootstock Selection plays a critical role in managing water scarcity. Drought-tolerant rootstocks such as Ramsey, 1103 Paulsen, and 140 Ruggeri are particularly valuable in warm climates as they are capable of accessing larger volumes of soil moisture due to their extensive root systems. In contrast, rootstocks like 101-14, which have high water requirements, are less suited to water-restricted environments. Effective soil preparation at planting is also crucial, as it promotes the development of a robust root system capable of supporting the vine under limited water conditions. Understanding the drought tolerance of rootstocks enables vineyard managers to make informed decisions during water budget planning, ensuring better resilience during dry seasons.

Irrigation Strategy during water-scarce periods is another key consideration. During the 2007–2009 drought in Australia's Mallee region, grape producers employed two main irrigation approaches: frequent short irrigations or less frequent deeper irrigations. The latter proved more efficient, as it minimized water loss to evaporation and ensured that a larger portion of the root zone had access to water. Shallow applications risk water remaining near the soil surface, where it is more prone to evaporation, particularly when using low-level sprinklers. Deep, well-monitored irrigations ensure water infiltrates to the root zone, optimizing its availability for vine uptake without causing deep drainage.

The effective management of irrigation through RDI, SDI, rootstock selection, and tailored irrigation strategies is essential for vineyards in regions experiencing water scarcity. These approaches not only help conserve water but also maintain vine health and grape quality, contributing to the sustainability of viticulture in challenging climatic conditions.

Water Saving Practices in Vineyards

Water conservation in vineyards is a critical component of sustainable viticulture, particularly in regions experiencing water scarcity or drought. Effective strategies involve immediate and long-term actions to optimize water use, maintain vine health, and ensure sustainable production.

Immediate strategies:

- **Water Budgeting:** Developing a water budget is a fundamental step in managing limited water resources. By estimating monthly water requirements using historical irrigation records and average water use data, vineyard managers can prioritize water allocation. This ensures that critical areas or varieties receive adequate irrigation, while lower-priority patches are managed accordingly. A drought management plan helps to streamline this process and allows for informed decision-making during periods of low water availability.

- **Irrigation Scheduling Devices:** Accurate soil moisture monitoring is essential to schedule irrigations effectively. Tools like tensiometers, though inexpensive and user-friendly, provide basic insights into soil moisture. However, more advanced devices, such as those that continuously log soil moisture data, offer superior accuracy and allow for precise determination of irrigation depth. These tools facilitate confident and informed decisions, enabling better water management. Installing multiple sensor depths helps determine the active root zone, further enhancing irrigation efficiency.

- **System Maintenance and Efficiency:** Regular maintenance of the irrigation system is critical to prevent water loss. Checking for leaks, blockages, and inaccurate water meter readings ensures that water is distributed evenly across the vineyard. Irrigation consultants can provide advice on improving system uniformity if needed. Redirecting flush water to the vineyard instead of allowing it to run to waste is another effective practice, particularly during drought conditions.

- **Avoiding Leaching Losses:** Water lost below the root zone wastes valuable resources and reduces irrigation efficiency. Monitoring soil moisture levels and adjusting irrigation depth accordingly minimizes leaching. Soil salinity sampling is also recommended, as water with elevated salinity during drought conditions may require strategic leaching programs to prevent salt accumulation.

- **Mulching:** Applying organic mulches, such as manure or slashed cover crops, over the wetted strip reduces soil evaporation and improves moisture retention. However, it is important to monitor soil moisture carefully, as certain mulches can initially act as a barrier to water penetration.

- **Night Irrigation:** Irrigating during nighttime significantly reduces evaporation losses compared to daytime irrigation. Low-level sprinkler systems, in particular, can save up to 20–30% of water when used at night. The savings with drip systems are variable but can be enhanced by the use of mulch, leaf litter, and shade over the wetted soil.

- **Preventing Runoff:** Runoff wastes water and reduces soil water availability. Ensuring proper soil structure through aeration and breaking up crusted soil surfaces enhances water penetration. Pulse irrigation, where water is applied intermittently (e.g., one hour on, one hour off), can further improve water infiltration and aeration.

- **Reducing Windbreak Irrigation:** Ceasing irrigation for windbreaks can conserve water, though long-term implications for vineyard protection must be considered. Where necessary, measures such as trenching or deep ripping to trim windbreak roots may prevent them from scavenging water from the vines.

- **Reusing Back-Flush Water:** Back-flush water from irrigation filters can be recycled by directing it to a settling tank. While the savings may be modest, every drop counts during water scarcity.

- **Reducing Transpiration:** Products like kaolin-clay-based foliar sprays claim to reduce water loss through transpiration, but their effectiveness may vary. Testing these products on a small scale before widespread application helps determine their utility.

- **Buying or Trading Water:** Purchasing or leasing additional water is a viable option when available. Weighing the cost of water against potential losses in crop value can guide decision-making. Tools and financial counsellors can assist in evaluating the feasibility of this approach.

- **Weed Management:** Eliminating weeds and cover crops reduces competition for water. Applying herbicides to create a protective mulch layer helps retain soil moisture and reduces evaporation.

- **Staying Informed:** Regular updates on water allocations, storage levels, and long-term weather forecasts allow vineyard managers to plan more effectively. Accessing resources like water authority websites and apps ensures that decisions are based on the latest information.

Long-Term strategies:

- **Patch-Specific Water Management:** Installing valves for individual vineyard patches allows for tailored irrigation based on canopy size, variety, and rootstock. This enables more precise matching of water applications to vine requirements, improving overall efficiency.

- **Sophisticated Scheduling Equipment:** Capacitance probes and other advanced scheduling devices provide highly accurate soil moisture data. These tools help avoid leaching losses and ensure effective irrigation, even during periods of limited water availability.

- **Drip Irrigation Systems:** Drip irrigation is one of the most water-efficient methods available. While the initial investment in a permanent drip system can be significant, the long-term benefits in water savings and vine health often justify the cost. Proper design and professional installation are critical for success. Transitioning from other systems to drip irrigation requires a period of adjustment for the root system, so careful management is necessary during this phase.

Creating a Water Budget for Small-Scale Wine Grape Growing

A water budget is a vital tool for managing irrigation in small-scale vineyards, particularly in areas with limited water availability or seasonal variability. It enables growers to optimize water use, ensure vine health, and maintain grape quality while conserving resources. Below is a guide on creating a water budget for small-scale wine grape production.

1: Assess Vineyard Water Requirements

Understanding the water needs of your grapevines is the foundation of a water budget. Water requirements depend on several factors:

- **Growth Stages:** Grapevines have varying water needs throughout their growth cycle. Key stages include:

  o **Budburst to Flowering:** Low water demand; adequate moisture is essential for early canopy development.

  o **Flowering to Fruit Set:** High water sensitivity; adequate water ensures successful fruit set.

  o **Fruit Set to Veraison:** Moderate to high demand; water deficits can reduce berry size for winegrapes.

  o **Veraison to Harvest:** Moderate demand; excessive water can dilute grape quality.

  o **Post-Harvest to Leaf Fall:** Low demand; adequate water supports carbohydrate storage for the next season.

- **Varietal Requirements:** Different grape varieties have unique water needs based on growth habits and fruiting characteristics.

- **Rootstock:** Drought-tolerant rootstocks require less irrigation than water-intensive ones.

- **Climate:** Local weather patterns, including temperature, rainfall, and evaporation rates, influence water needs.

- **Soil Type:** Soil texture and depth affect water-holding capacity. Sandy soils drain quickly, while clay soils retain moisture longer.

2: Determine Water Availability

Assess the total water resources available for irrigation. This includes:

- **Stored Water:** Calculate the volume of water stored in tanks, reservoirs, or ponds.

- **Rainfall:** Use historical weather data or rain gauges to estimate average rainfall during the growing season.

- **Groundwater:** If using bore water, ensure sustainable extraction rates and account for quality considerations like salinity.

3: Measure Vineyard Area and Layout

Determine the vineyard size and layout to calculate the total irrigated area. For small-scale vineyards, measure the row spacing, vine spacing, and the number of vines to estimate water requirements accurately.

4: Estimate Vine Water Use

Vine water use is typically measured as crop evapotranspiration (ETc), which combines water lost through vine transpiration and soil evaporation. To calculate ETc:

- Obtain local reference evapotranspiration (ETo) data from weather stations or online resources.

- Adjust ETo using a crop coefficient (Kc), which varies by vine growth stage. For wine grapes:

  - Early season: Kc = 0.2–0.3

  - Mid-season: Kc = 0.7–0.8

  - Late season: Kc = 0.5–0.6

2. Multiply ETo by Kc to estimate ETc for your vineyard.

5: Develop an Irrigation Schedule

Based on ETc calculations and soil moisture data, create an irrigation schedule:

- **Frequency:** Determine how often vines need irrigation based on soil type, weather, and growth stage.

- **Depth:** Calculate how much water to apply per irrigation event. Ensure water penetrates the root zone without causing runoff or leaching.

6: Monitor Soil Moisture

Use tools like tensiometers, soil moisture probes, or capacitance sensors to monitor soil moisture levels. These devices help ensure irrigation is precise and prevents over- or under-watering.

7: Adjust for Efficiency and Losses

Account for irrigation system efficiency when planning water applications. Drip systems, for example, have efficiencies of 85–95%, meaning some water may be lost to evaporation or leaks. Adjust your water budget to compensate for these losses.

8: Prioritize Water Allocation

In times of water scarcity, prioritize water use based on:

- **High-value Varieties:** Focus on grape varieties critical for wine production.

- **Key Growth Stages:** Allocate more water during flowering, fruit set, and veraison to optimize yield and quality.

- **Healthy Patches:** Favor productive vineyard sections over struggling or low-yield areas.

9: Record and Review

Maintain detailed records of water use, rainfall, and vine performance. Regularly review and adjust the water budget based on actual conditions and vineyard response.

Step 10: Plan for Long-Term Sustainability

Incorporate long-term strategies into your water budget, such as:

- **Rainwater Harvesting:** Capture and store rainwater for future irrigation.

- **Soil Health Improvement:** Use mulches and organic matter to enhance soil moisture retention.

- **Efficient Irrigation Systems:** Invest in drip or subsurface irrigation for precision water delivery.

Creating a water budget for small-scale wine grape growing ensures efficient water use, supports vine health, and maintains grape quality. Assessing water needs, measuring available resources, and implementing precise irrigation schedules, allows growers to achieve sustainable viticulture practices while navigating the challenges of water scarcity and climate variability.

Managing Cover Crops and Ground Cover

Cover crops play a vital role in vineyard management, serving multiple ecological and agronomic purposes. They are non-economic crops cultivated primarily to enhance soil health, prevent erosion, suppress weeds, and contribute organic matter, which in turn improves soil fertility. The incorporation of cover crops in vineyards has been shown to effectively protect the soil from erosion caused by rainfall, as they reduce the impact of raindrops and help maintain soil structure [318, 319]. Additionally, cover crops can facilitate vineyard access during wet weather, which is crucial for vineyard operations [320].

The selection of appropriate cover crop species is essential for maximizing their benefits while minimizing competition with grapevines for water and nutrients. Certain species, particularly legumes, can enhance soil fertility through nitrogen fixation, while others can suppress weed growth by outcompeting them [321]. However, the management of cover crops must be carefully tailored to the specific vineyard conditions, including soil type, slope, and vine age, to avoid negative impacts on grapevine vigour and yield [322, 323]. For example, while perennial cover crops can be beneficial, they may also restrict water access in drier years, necessitating a balance between cover crop benefits and potential drawbacks [324].

Moreover, the role of cover crops in integrated pest management is increasingly recognized. By promoting biodiversity, cover crops can support natural pest predators and enhance the overall resilience of the vineyard ecosystem [322]. This biodiversity can lead to improved pest control, reducing the reliance on chemical pesticides, which is a significant advantage for sustainable viticulture [318, 319]. However, it is important to note that environmental variability can influence the effectiveness of cover crops; what may work well in one season might not be effective in another due to changing climatic conditions [325, 326].

Climate change poses additional challenges for vineyard management, as increased rainfall intensity and shifting weather patterns can affect soil moisture and nutrient availability [325, 326]. Vineyard managers are encouraged to adopt adaptive management strategies, including the use of cover crops, to mitigate these impacts and enhance the resilience of their operations [327]. The careful management of cover crops, including practices like mowing or rolling, can also help regulate soil and canopy temperatures, which is crucial for frost prevention and mitigating heat stress during the growing season [324, 328].

Cover crops provide numerous benefits in vineyard management, significantly contributing to sustainable practices and long-term soil health. One of their primary advantages is erosion control. Cover crops stabilize the soil, reducing the risk of erosion caused by water runoff or wind. Their robust root systems anchor the soil in place, while their foliage shields the ground from raindrop impact, which can dislodge soil particles. This is particularly critical in sloped vineyards where erosion risk is high.

Another key benefit of cover crops is their contribution to organic matter and soil structure. As cover crops decompose, they add organic material to the soil, enhancing its structure and improving water infiltration and retention. This process creates a more favourable environment for root growth and supports the activity of beneficial soil microorganisms.

Figure 31: Cover cropping at Granton Vineyard (fava beans). Stefano Lubiana, CC BY 2.0, via Flickr.

Legume cover crops, such as clover and vetch, offer the additional advantage of nitrogen fixation. These plants establish a symbiotic relationship with nitrogen-fixing bacteria, enriching the soil with nitrogen. This natural process reduces the reliance on synthetic fertilizers and supports healthier vine growth.

Cover crops also play a vital role in recycling or scavenging unused nutrients. Deep-rooted species capture nutrients that might otherwise leach out of the soil and become inaccessible to vines. When these crops are mowed or decompose, they release these nutrients back into the soil, making them available for grapevine uptake and improving nutrient cycling.

In vineyards with vigorous vine growth, cover crops help regulate vine growth by competing for water and nutrients. This controlled competition curbs excessive vine vigour, balancing canopy development and enhancing fruit quality. Certain cover crops also attract beneficial insects, contributing to pest control. Flowering species like buckwheat and alyssum provide habitats for predatory insects that naturally manage vine pests, reducing the need for chemical interventions.

Weed suppression is another significant benefit of cover crops. Their rapid growth and dense coverage outcompete weeds, forming a natural barrier that inhibits weed seed germination and growth. This reduces the need for herbicides and supports a more sustainable vineyard ecosystem.

Cover crops enhance soil microbial activity by providing organic material and maintaining soil moisture. This stimulates beneficial microbial processes, which improve nutrient availability and help suppress soil-borne diseases. They also assist in soil water management by improving water infiltration and reducing evaporation. However, careful management is required to ensure that cover crops do not compete excessively with vines for water.

Lastly, cover crops regulate soil temperature, providing insulation that minimizes temperature fluctuations. They help maintain cooler soil in hot climates, protecting vine roots from heat stress, and reduce frost risk by moderating soil cooling in colder conditions. These benefits collectively make cover crops a cornerstone of sustainable viticulture.

While cover crops offer significant benefits in vineyard management, they also come with certain drawbacks that require careful consideration. One of the primary challenges is the additional costs associated with establishing and maintaining cover crops. Expenses for seeds, planting, mowing, and potential irrigation can add up, and these costs may not always be immediately offset by the benefits they provide. For small-scale or resource-limited vineyards, these financial implications can be a significant concern.

Pest problems are another potential drawback. Some cover crops can act as hosts for pests like nematodes or rodents, which may subsequently attack the grapevines. This risk underscores the importance of selecting pest-resistant or pest-neutral cover crop species and employing integrated pest management practices to minimize the threat to the vineyard.

The risk of frost is another factor to consider. Cover crops can alter air movement and soil temperature, which, in certain conditions, may increase the likelihood of frost damage. Effective management strategies, such as mowing the cover crops before frost-prone periods, are necessary to mitigate this risk while maintaining the benefits of soil protection and structure.

The life cycle of cover crops further influences their management and suitability for vineyard systems. Annual cover crops complete their life cycle within a single growing season, requiring replanting each year. These crops are often chosen for specific short-term benefits like nitrogen fixation or erosion control. However, the need for annual replanting adds labour and costs.

Perennial cover crops, on the other hand, persist for several years, reducing the labour and expense associated with replanting. While this longevity can be an advantage, these crops may require more intensive management to prevent excessive competition with grapevines for water and nutrients. Their persistence also demands a strategic approach to mowing and maintenance to ensure they support, rather than hinder, vineyard productivity.

Biennial cover crops, which live for two years, provide a middle ground between annual and perennial options. They offer extended soil cover and fertility benefits without the need for annual replanting. However, their use requires careful planning to align their life cycle with vineyard management goals and seasonal needs.

Understanding and addressing these drawbacks is essential to effectively integrate cover crops into vineyard systems. With thoughtful planning, the challenges associated with cover crops can be mitigated, allowing vineyards to maximize their benefits while minimizing their potential negative impacts.

Different types of cover crops play distinct roles in vineyard management, each offering unique benefits to the soil and vine health. Among these, legumes, grasses, brassicas, and buckwheat are commonly used, each selected based on specific vineyard needs and seasonal conditions.

Legumes, such as clover and vetch, are valued for their ability to fix atmospheric nitrogen into the soil through a symbiotic relationship with nitrogen-fixing bacteria. This natural enrichment reduces the need for synthetic fertilizers, supporting sustainable vineyard practices. Cool-season legumes thrive during colder months, while warm-season varieties flourish in warmer conditions, offering flexibility in their application. With their low carbon-to-nitrogen ratio, legumes decompose quickly, releasing nutrients back into the soil to enhance vine growth.

Grasses, including species like rye and oats, are recognized for their contribution to building soil structure. Their high carbon-to-nitrogen ratio ensures they decompose more slowly, creating long-lasting organic matter that improves soil texture and water retention. Cool-season grasses are typically planted in the fall to provide cover through the winter, while warm-season grasses are sown in the summer to support soil health during the active growing season.

Brassicas, such as radishes and mustards, serve dual purposes in vineyard management. Their deep-rooted systems help to aerate and break up compacted soils, improving water infiltration and root penetration. Additionally, some brassicas release allelopathic compounds that suppress weed growth, providing a natural alternative to chemical herbicides. These attributes make them an effective choice for maintaining soil health and weed control in vineyards.

Buckwheat is a fast-growing cover crop often used in vineyards with short growing windows. Its rapid growth suppresses weeds and reduces competition for resources. Additionally, buckwheat attracts beneficial insects like pollinators and pest predators, enhancing biodiversity within the vineyard ecosystem. Its adaptability and quick turnover make it a versatile option for cover cropping between vine growth cycles.

By carefully selecting the type of cover crop to match vineyard goals and environmental conditions, growers can optimize soil health, manage pests and weeds, and support sustainable viticulture practices. Each type of cover crop offers specific advantages, and their integration into vineyard management plans requires a strategic approach to maximize their benefits.

Figure 32: Newly planted Opus One Vineyard in the Napa Valley of California. This picture shows a close up of the irrigation piping, staking, wire trellising and cover crop between the vines that goes into establishing a vineyard. Helder Ribeiro, CC BY-SA 2.0, via Wikimedia Commons.

Deciding between monocultures and mixtures for cover cropping in vineyards involves weighing the simplicity and targeted benefits of single-species planting against the ecological advantages of diverse species combinations. Each approach offers unique benefits and challenges, making the choice dependent on specific vineyard goals and conditions.

Planting a monoculture involves growing a single cover crop species, which simplifies management and allows growers to focus on achieving targeted outcomes. For example, legumes are often chosen for their nitrogen-fixing ability, while grasses might be selected to control erosion. Monocultures are easier to establish and manage, as all plants share similar growth habits and nutrient requirements. However, they may lack the biodiversity needed to support pest control and soil health comprehensively.

Mixtures of cover crops combine different species to provide multiple benefits simultaneously, such as nutrient cycling, pest control, soil stabilization, and increased biodiversity. These combinations create a more resilient vineyard ecosystem by fostering a variety of root structures, nutrient contributions, and habitats for beneficial insects. While mixtures offer a broader range of advantages, they require more careful planning and management to ensure compatibility among species and to avoid competition with grapevines.

When managing cover crops, the decision between tilled and no-till systems further influences vineyard health and sustainability. In annually tilled systems, annual cover crops are incorporated into the soil to add organic matter and cycle nutrients. Although effective, this practice requires more labour and disturbs soil structure, potentially leading to erosion and compaction.

No-till cover cropping systems, on the other hand, preserve soil integrity by leaving cover crops on the surface as mulch. This reduces erosion, improves water retention, and supports microbial activity. However, suppression of cover crop growth in no-till systems may require mowing or herbicide application, depending on the vineyard's management strategy.

For annual cover crops in non-tillage systems, mowing the crops and leaving the residue as mulch provides organic matter and retains soil moisture without disturbing the soil. In contrast, perennial cover crops grow continuously and offer long-term soil cover, requiring less labour but necessitating careful management to prevent competition with grapevines.

Cover crops also play a role in pest management. Certain species attract beneficial insects that prey on vineyard pests, reducing the need for chemical interventions. Managing flowering cycles ensures these insects have consistent habitats. Cover crops also improve soil health and drainage, which reduces conditions favourable for fungal diseases. For weed suppression, fast-growing cover crops compete effectively for resources and shade the soil, inhibiting weed germination. Residual mulch from these crops further enhances weed control.

When selecting cover crop species, several factors must be considered, including vineyard vigour, frost risk, nitrogen contribution, soil erosion control, moisture availability, and weed management objectives. Cost and the specific benefits desired also guide the decision. Proper establishment and management are essential for successful cover cropping. Using seeds from reputable sources and planting at the right time ensures maximum benefits, with fall planting ideal for cool-season crops and spring planting suitable for warm-season varieties. Preparing the seed bed ensures good seed-to-soil contact for germination, while seeding methods like broadcast seeders, seed drills, or no-till drills minimize soil disturbance.

Supplementing deficient nutrients with fertilizers may be necessary for optimal cover crop performance. Control methods such as mowing, herbicides, or tillage can manage cover crop growth and prevent competition with grapevines, ensuring a balanced vineyard ecosystem. By thoughtfully integrating these practices, growers can maximize the benefits of cover crops while mitigating potential drawbacks.

Chapter 6
Pruning and Training for Quality and Yield

Winter and Summer Pruning

Pruning grapevines is an essential practice in vineyard management that directly affects vine health, fruit quality, and productivity. Grapevines require two main pruning periods: winter pruning, performed during dormancy, and summer pruning, conducted during the active growing season. Each serves a unique purpose and is tailored to the vine's growth cycle and production goals.

Winter pruning and summer pruning are distinct vineyard management practices, differing in timing, purpose, and intensity, each playing a crucial role in grapevine health and productivity.

Timing is one of the key distinctions between the two. Winter pruning takes place during the vine's dormancy, typically from late fall to early spring. This allows growers to shape the vine and prepare it for the upcoming growing season. In contrast, summer pruning is conducted during the active growing phase, from late spring to early summer, when the vine is developing shoots, leaves, and fruit.

The purpose of these pruning methods also varies. Winter pruning focuses on establishing the vine's structure and regulating its yield potential by determining how many buds are left to grow into shoots and grape clusters. On the other hand, summer pruning aims to optimize the growing environment, focusing on improving light penetration, air circulation, and fruit quality. This includes practices like shoot thinning, leaf removal, and hedging.

Intensity further sets these practices apart. Winter pruning is more structural and severe, involving the removal of large portions of the previous year's growth to prepare the vine for balanced production. Summer pruning, however, is lighter and corrective, targeting excessive or poorly positioned growth to fine-tune the canopy and enhance the conditions for fruit ripening.

Together, winter and summer pruning complement each other, ensuring a well-balanced vine that supports sustainable yields and high-quality grape production.

First Growing Season: When planting a grapevine for the first time, the primary focus should be on allowing the plant to establish a robust root system. During this initial year, let the vine grow freely without pruning to enable its roots to develop deeply and spread effectively. At the end of this first growing season, select the straightest and sturdiest shoot from the growth and train it vertically using a post or bamboo stake as a guide. Secure the shoot gently with twine or plastic tape to encourage upright growth. Remove all lower shoots along the trunk, focusing the vine's energy on the chosen shoot. Cut this selected shoot back to about three to four buds during the first winter, preparing it for more structured growth in the following season.

Second Growing Season: In the second year, the goal is to continue elongating the trunk and start forming the vine's structure. Choose the strongest and most upright shoot for the trunk, and tie it to the post for support. If the vineyard layout involves two-tier growth, begin training the lower layer by selecting two shoots at approximately 30 inches above the ground on either side of the trunk. Tie these shoots to the support wire, as they will form the vine's cordons. Remove any additional shoots along the trunk that are not part of the desired structure. As the trunk reaches the top support wire, pinch the tip of the central shoot to encourage lateral growth and development of side shoots.

Second Winter: During the second winter, refine the vine's structure by removing unwanted shoots along the trunk and arms (cordons). This ensures the energy is directed toward the selected framework for fruiting and structural development.

Third Summer: By the third summer, the focus shifts to maintaining the vine's structure and encouraging proper shoot development. Remove any shoots sprouting along the trunk and allow the trained side shoots to grow and establish. The subsequent winter will be crucial for deciding the pruning method—spur pruning or cane pruning—depending on the grape variety and vineyard goals.

Winter Pruning

Winter pruning is a critical horticultural practice performed during the grapevine's dormancy period, typically from late fall to early spring. This technique is essential for shaping the vine, regulating fruit production, and promoting healthy growth for the upcoming season. The primary purpose of winter pruning is to remove the previous season's growth, which prepares the vine for balanced growth and optimal fruit production. By determining the number of buds that will produce shoots, leaves, and grape clusters, winter pruning directly influences the yield and quality of the grapes produced [329, 330].

Figure 33: Winter pruning grapevines at Stefano Lubiana. Mark Smith, CC BY 2.0, via Wikimedia Commons.

The benefits of winter pruning are multifaceted. It not only enhances the vine's structure but also plays a significant role in managing vine health by removing dead, diseased, or weak wood, which can harbor pests and diseases [331]. Additionally, careful pruning helps to balance the vine's vigour by leaving an appropriate number of buds, ensuring healthy growth without overburdening the plant [332]. Research has shown that the timing and intensity of pruning can significantly affect vine yield and grape quality, with studies indicating that delayed pruning can lead to fluctuations in yield due to variations in bud fertility and berry weight [329, 330, 332].

Two primary techniques are commonly employed in winter pruning: cane pruning and spur pruning. Cane pruning involves selecting one or two healthy canes from the previous year's growth and trimming them to retain a specific number of buds, typically between 8 and 12, depending on the vine's vigour and variety. This method is particularly prevalent in cooler climates and is favoured for high-quality grape production [333]. Conversely, spur pruning entails cutting back canes to short spurs, each containing 2 to 3 buds. This technique is generally used in warmer climates and for grape varieties that reliably produce crops from shorter wood [334, 335]. Both methods may incorporate a renewal spur, which is a short shoot left near the vine's main structure to develop into new fruiting wood for the following year [336].

When considering winter pruning, several important factors must be taken into account. The selection of moderate vigour dormant shoots is crucial for forming next season's canes and spurs [333]. Moreover, the vineyard's trellis system and the desired grape quality should guide the pruning intensity [330, 337]. Research indicates that pruning performed at specific stages, such as the bud swollen stage, can help mitigate risks associated with late spring frost events, thereby enhancing overall vine performance [336, 337]. Furthermore, the application of pruning techniques must adapt to the unique conditions of each vineyard, as factors such as vine crop load, berry size, and water status can significantly influence grape ripening and overall yield [330, 331, 338].

Spur Pruning

Selecting Spurs: For vines suitable for spur pruning, it is essential to select upright and well-placed spurs along the cordons. Spurs should be evenly spaced, roughly 6 inches apart, with about six to seven spurs on each half of the cordon. Shorten the chosen spurs to two buds, ensuring a productive yet manageable growth pattern for the coming season.

Varieties That Respond Well to Spur Pruning: Grape varieties such as Flame, Black Monukka, Ruby, Muscat of Alexandria, Chardonnay, Merlot, Zinfandel, Pinot Noir, Red Globe, and Shiraz are ideal candidates for spur pruning. Their growth habits and fruiting patterns are well-suited to this method, ensuring balanced production and ease of maintenance.

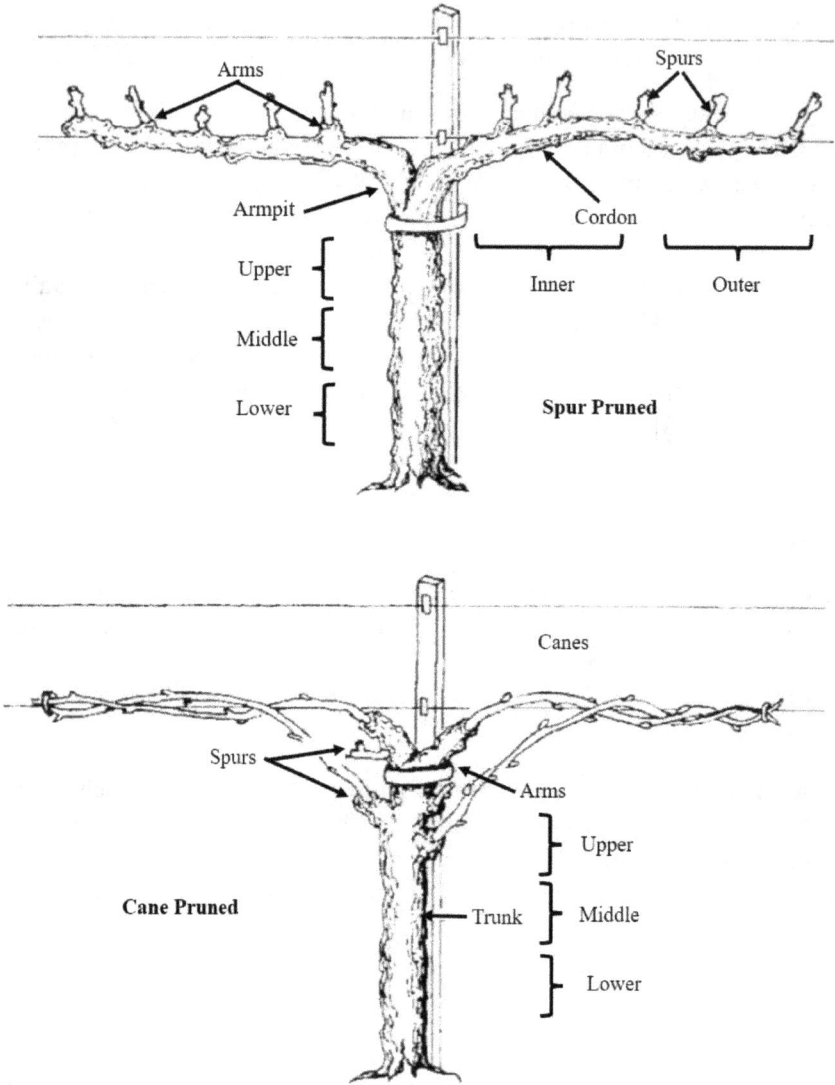

Figure 34: Spur pruning and cane pruning.

Cane Pruning

Selecting Fruiting Canes and Renewal Spurs: In late winter or early spring, evaluate the previous year's growth to identify one-year-old canes for fruiting. Choose two canes on each side of the trunk that are strong, exposed to sunlight, and close to the trunk. Ideal fruiting canes should have a diameter of ¼–½ inch and short internodes, approximately 3–4 inches apart. Save two additional canes near the base of the fruiting canes as renewal spurs for the following year.

Trim the selected fruiting canes to 10–15 buds and tie them to the guide wire or trellis. The renewal spurs should be pruned back to two buds. This cycle ensures continuous productivity while maintaining the vine's structure.

Varieties That Respond Well to Cane Pruning: Grape varieties like Cabernet Sauvignon, Chardonnay, Thompson, Flame, Black Monukka, Ruby, Red Globe, Himrod, Sauvignon Blanc, Shiraz, Merlot, and Zinfandel benefit from cane pruning. This method encourages consistent yield and optimal fruit quality for these cultivars.

Mystery Grapes: How to Decide the Pruning Method: When dealing with neglected or unidentified grapevines, start by cane pruning during the first dormant season. Monitor where blooms appear during the growing season. If the blooms are concentrated near the main trunk, switch to spur pruning the following year. Conversely, if blooms occur along the entire length of the cane, continue with cane pruning. Observing the vine's fruiting habits provides valuable insights into the most effective pruning approach for maximizing yield and grape quality.

Summer Pruning

Summer pruning, also referred to as canopy management, is a critical practice in viticulture that occurs during the growing season, typically from late spring to early summer. This technique is employed to control excessive vine growth, enhance light penetration, and improve air circulation within the canopy, which are essential for maintaining vine health and optimizing grape quality [339-341]. By managing the canopy effectively, growers can create a microclimate that supports disease prevention, efficient photosynthesis, and optimal fruit ripening [339, 341].

The benefits of summer pruning are multifaceted. It helps in managing vine vigour, directing the plant's energy towards the development of high-quality fruit, and optimizing the microclimate around grape clusters. This is particularly important for preventing diseases such as powdery mildew and botrytis, which thrive in overly dense canopies [342, 343]. Techniques such as shoot thinning, shoot positioning, leaf removal, topping or hedging, and lateral removal are commonly employed to achieve these goals [340, 344]. For instance, shoot thinning involves the removal of excess or poorly positioned shoots early in the season to balance the canopy and ensure adequate light and air circulation [343, 345]. Similarly, leaf removal around grape clusters enhances sun exposure and air movement, which aids in ripening and reduces disease risk [166, 346, 347].

However, careful consideration is necessary when implementing these techniques. Excessive leaf removal can expose grape clusters to sunburn, particularly in hot climates, and timing is crucial; for example, leaf removal is best performed post-fruit set to avoid overexposure during critical growth stages [339, 346, 348]. Balancing the vine's canopy is essential to ensure sufficient leaves for photosynthesis while preventing excessive shading of the fruit zone [349]. Techniques such as shoot positioning and lateral removal can help maintain an open canopy, facilitating vineyard operations like spraying and harvesting while also optimizing light exposure for the grapes [340, 345, 350].

Training Systems for Small-Scale Vineyards

The use of vine training systems in viticulture serves primarily to assist in canopy management, aiming to achieve a balance between adequate foliage for photosynthesis and minimizing excessive shading. Properly managed canopies ensure optimal grape ripening and reduce the risk of diseases. Additionally, specific training systems can regulate yields and support mechanized tasks like pruning, irrigation, spraying pesticides or fertilizers, and harvesting.

When selecting a training system, growers must consider the vineyard's climate, as sunlight, humidity, and wind significantly influence the effectiveness of a system. For instance, expansive canopies, like those in the Geneva Double Curtain system, are ideal for promoting photosynthesis but may offer little wind protection. In regions like Châteauneuf-du-Pape, where the mistral winds are strong, compact and protective systems are preferred to shield the fruit.

While often used interchangeably, trellising, pruning, and vine training are distinct practices. Trellising refers to the physical support structures, including stakes, posts, and wires, used to guide vine growth. Pruning involves cutting and shaping the vine, particularly during the dormant season, to determine the number of buds that will produce grape clusters. Summer pruning involves removing excess shoots and clusters to improve airflow and sunlight penetration. Vine training systems combine these practices to manage the canopy, influencing yield and grape quality by optimizing light exposure and reducing disease risks.

The practice of training vines dates back to ancient civilizations like the Egyptians and Phoenicians, who discovered that specific techniques improved yields. The Greeks, during their colonization of southern Italy, introduced vine-staking methods, earning the region the name Oenotria, or "land of staked vines." Roman writers like Columella and Pliny the Elder offered insights into effective vine training.

Traditions guided vine training practices in many regions until the 20th century when wine laws codified these methods, such as the French AOC system. In the 1960s, New World regions like California, Australia, and New Zealand began researching vine training systems, adapting them to local conditions and modern winemaking styles. These studies have continued into the 21st century, leading to innovative systems tailored to specific mesoclimates and labour needs.

The primary goal of vine training systems is canopy management, ensuring sufficient sunlight and airflow while preventing excessive shading. Grapevines, as climbing plants, require support to prevent their canopy and clusters from sagging to the ground. Allowing vines to touch the ground increases the risk of root suckering, which can introduce pests like phylloxera and diminish the vine's resources.

Vine training systems also enhance labour efficiency and mechanization. By keeping the fruiting zone at an accessible height, vineyard tasks like pruning, spraying, and harvesting become less labour-intensive. Moreover, these systems prevent excessive shading, which can hinder photosynthesis and reduce grape quality. Excessive shade decreases sugar and anthocyanin levels while increasing potassium and malic acid, negatively impacting wine quality. Limited airflow from dense canopies can also promote fungal diseases like powdery mildew and gray rot.

A grapevine's canopy includes its trunk, cordons, stems, leaves, flowers, and fruit. Training focuses primarily on the woody structures—cordons and fruiting canes. Cane-trained systems involve pruning the vine back to a spur in winter, leaving a single cane for the next season's growth. Examples include the Guyot and Pendelbogen systems. Spur-trained systems retain permanent cordons, with only excess buds pruned. Examples include goblet (bush vines) and Cordon de Royat.

Figure 35: Typical Guyot is involves cane training where the left and right "arms" are 1 year old canes that get removed at the end of the year. Mondavi's system is a modification of Guyot's which combines the Cordon de Royat practice of keeping a two permanent spurs (the left and right "arms" of the T) and allowing new canes to grow upwards and fruit. After harvest these canes will be removed, leaving only the "T-shape" to continue to age. Jason DeRusha, CC BY-SA 2.0, via Wikimedia Commons.

Typical Guyot is involves cane training where the left and right "arms" are 1 year old canes that get removed at the end of the year. Mondavi's system is a modification of Guyot's which combines the Cordon de Royat practice of keeping a two permanent spurs (the left and right "arms" of the T) and allowing new canes to grow upwards and fruit. After harvest these canes will be removed, leaving only the "T-shape" to continue to age. Jason DeRusha, CC BY-SA 2.0, via Wikimedia Commons.

Cordons are trained unilaterally or bilaterally, extending horizontally along wires in systems like Lyre or Scott Henry. Alternatively, vertical trellising systems like VSP position fruiting canes upwards. Certain systems require unique techniques, such as arching canes in the Pendelbogen or directing them downward in the Scott Henry system.

Vine training systems are classified by trunk height (high or low-trained), pruning method (cane or spur), and canopy style (free or wire-constrained). High-trained systems, such as the Lenz Moser system, offer better frost protection, while low-trained systems like gobelet are closer to the ground. Systems like Guyot and cordons can be adapted for high or low training.

Further distinctions include trellising configurations, such as the number of wires and whether they are movable. Movable wires, like those in the Scott Henry system, allow growers to adjust the canopy's size and structure. In regions prone to wind damage, temporary or permanent stakes provide additional support.

Balancing Growth and Fruit Production

Achieving a balance between vine growth and fruit production is critical for the cultivation of high-quality wine grapes. This balance ensures that the grapevine allocates its resources effectively, which is essential for both vegetative growth and fruit development. The equilibrium between these two aspects directly impacts vine health, fruit composition, and ultimately, the quality of the wine produced [351, 352].

A well-balanced grapevine features a healthy canopy that supports adequate photosynthesis, which is vital for energy production necessary for vine growth and fruit ripening. Excessive vegetative growth, often termed "over-vigorous" growth, can lead to shading within the canopy. This shading reduces sunlight exposure for grape clusters, hindering the development of sugars and other compounds important for flavour and aroma [353, 354]. Conversely, insufficient vegetative growth, or "under-vigour," can compromise the vine's ability to ripen fruit effectively and recover post-harvest [355, 356]. Thus, maintaining an optimal balance is essential for maximizing both yield and quality.

Several factors influence the balance between growth and fruit production, including soil fertility, water availability, climate, vineyard management practices, and vine training systems. Fertile soils and adequate water supply can promote vigorous growth, necessitating careful management to prevent shading and competition for resources [357, 358]. Conversely, poor soil conditions or drought stress can limit growth, adversely affecting the vine's capacity to sustain a healthy canopy and ripen fruit [252, 351]. Additionally, the choice of grapevine variety and rootstock significantly impacts growth patterns; some varieties are naturally more vigorous, while others are more restrained, which can affect overall vineyard management strategies [359, 360].

Effective canopy management practices, such as pruning, shoot thinning, and leaf removal, are essential for achieving the desired balance between growth and fruit production. Winter pruning determines the number of buds that will produce shoots and clusters, directly influencing vine vigour and crop load [361]. During the growing season, practices like shoot thinning and leaf removal enhance sunlight penetration and airflow within the canopy, reducing disease risk and improving fruit quality [362, 363]. Properly managed canopies maintain the photosynthetic efficiency necessary for ripening grapes without creating excessive shading [354, 364].

Crop load, defined as the ratio of fruit to vine canopy, is a critical aspect of maintaining balance. An excessive crop load can overwhelm the vine's resources, leading to delayed ripening and lower sugar levels, which negatively impacts wine quality [351, 360]. Techniques such as green harvesting, which involves removing excess fruit clusters during the growing season, are employed to adjust crop load, ensuring that the remaining clusters receive adequate resources for optimal ripening [252, 365]. Conversely, an insufficient crop load can lead to over-vigorous growth, wasting energy and reducing wine complexity [353, 356].

Balancing growth and fruit production is not solely about immediate outcomes; it also concerns the long-term health of the vineyard. Practices such as overcropping or undercropping can weaken vines over time, diminishing their productivity and longevity [252]. Sustainable vineyard practices, including regulated deficit irrigation (RDI) and the use of cover crops, can help maintain soil health and moderate vine growth over time, ensuring a sustainable approach to vineyard management [252, 358].

Reducing Stress on Vines for Longevity

Reducing stress on grapevines is essential for ensuring their long-term health and productivity. Stress can arise from a variety of sources, including environmental conditions, pests, diseases, and vineyard management practices. Minimizing stress helps vines maintain consistent yields, enhances fruit quality, and extends their productive lifespan.

Stress in grapevines occurs when unfavourable conditions disrupt their normal growth and physiological processes. While some stress, such as a mild water deficit, can be beneficial for improving fruit quality, excessive or prolonged stress weakens the vine, reducing vigour and yield potential. Common stressors include drought, extreme temperatures, nutrient deficiencies, pest and disease pressure, and improper pruning or training practices.

Proper water management is critical to minimize stress. Overwatering can lead to excessive vine vigour and increased disease risk, while underwatering can result in wilting, reduced photosynthesis, and poor fruit development. Employing strategies like regulated deficit irrigation ensures vines receive just enough water to sustain growth and ripen fruit without promoting excessive vegetative growth. Tools like tensiometers or soil moisture probes help monitor water levels and enable precise application.

Balanced soil nutrition is another key factor in reducing stress. Nutrient deficiencies weaken vines and make them more susceptible to environmental pressures. Regular soil and tissue testing identify nutrient imbalances, allowing targeted applications of fertilizers or organic amendments. However, over-fertilizing, particularly with nitrogen, should be avoided, as it can encourage excessive vegetative growth and increase susceptibility to pests and diseases.

Canopy management plays a vital role in optimizing sunlight exposure, airflow, and temperature regulation within the vine. Practices such as shoot thinning, leaf removal, and canopy positioning maintain a balanced structure. Improved airflow reduces humidity and the risk of fungal diseases, while adequate sunlight ensures efficient photosynthesis and proper fruit ripening.

Pest and disease control is crucial to reducing vine stress. Pests and diseases divert the vine's energy from growth and fruit production. Integrated Pest and Disease Management (IPDM) strategies, including biological controls, cultural practices, and selective chemical treatments, protect vines without adding unnecessary stress. Early detection and prompt intervention are critical to limiting damage.

Temperature regulation is another essential factor. Extreme heat or cold can severely stress grapevines. Mitigating heat stress involves using shade cloths, applying kaolin-based sprays, or maintaining a healthy canopy to shield vines from intense sunlight. Frost risks can be reduced with frost fans, sprinklers, or heaters in vulnerable areas. Site selection and vineyard design also play significant roles in protecting vines from temperature extremes.

Pruning and training practices directly affect stress levels. Over-pruning or improper training can disrupt vine balance, reducing its capacity to produce and ripen fruit. Dormant season pruning helps establish a manageable structure, while summer pruning removes excess growth. Training systems suited to the vineyard's climate and vine variety support healthy development and minimize mechanical stress.

Soil health is integral to vine resilience. Practices such as cover cropping, mulching, and reducing tillage maintain soil structure and prevent erosion. Adding compost or organic matter enhances fertility and water retention, alleviating environmental stress.

Mechanical damage from equipment used during pruning, spraying, or harvesting can stress vines if not managed carefully. Precision mechanization and well-trained labour reduce damage to trunks, shoots, and root zones, safeguarding vine health.

Managing crop load is essential to preventing overexertion. Overloading vines with excessive fruit diverts resources from healthy development and quality production. Practices like green harvesting, which removes excess clusters early in the season, focus the vine's energy on fewer, higher-quality grapes, promoting longevity.

Minimizing stress on grapevines enhances immediate productivity and quality while ensuring their long-term sustainability. Healthy vines recover better from environmental challenges, resist pests and diseases, and adapt to changing climatic conditions. This reduces the need for costly interventions and supports consistent yields over time. A proactive approach to stress management safeguards the economic viability of vineyards and contributes to producing high-quality wines year after year.

Chapter 7
Harvesting and Post-Harvest Practices

Determining Optimal Harvest Time

The harvest of wine grapes, known as vintage, is a pivotal stage in the winemaking process. The timing of the harvest is primarily influenced by the ripeness of the grapes, determined by sugar, acid, and tannin levels, and is guided by the winemaker's desired wine style. Weather conditions also play a significant role in shaping the harvest schedule, with factors such as heat, rain, hail, and frost posing risks to grape quality and potentially introducing vine diseases.

In addition to deciding when to harvest, winemakers and vineyard owners must choose between manual hand-picking and mechanical harvesting methods. Harvest season typically occurs between August and October in the Northern Hemisphere and February to April in the Southern Hemisphere. However, due to variations in climate, grape varieties, and wine styles, grape harvesting can take place at different times throughout the year, depending on the region. In New World wine regions, this crucial period is often referred to as "the crush."

Figure 36: Using a sickle knife to hand harvest some wine grapes. Hahn Family Wines, CC BY 2.0, via Wikimedia Commons.

The majority of the world's wine-producing regions are situated between the temperate latitudes of 30° and 50° in both the Northern and Southern Hemispheres. Regions closer to the equator typically experience earlier harvests due to their warmer climates. In the Northern Hemisphere, vineyards in Cyprus can begin harvesting as early as July, while in California, sparkling wine grapes are sometimes picked in late July to early August at a slightly unripe stage to preserve acidity. The peak harvest period for most Northern Hemisphere vineyards falls between late August and early October, though late-harvest wine grapes may be picked well into the autumn. In colder regions like Germany, Austria, the United States, and Canada, ice wine grapes are often harvested as late as January.

In the Southern Hemisphere, harvesting can begin as early as January 1 in warmer regions, such as parts of New South Wales, Australia. Most vineyards in the Southern Hemisphere harvest between February and April, though cooler areas, like Central

Otago in New Zealand, may extend their harvest for late-harvest wine grapes into June.

Recent climate changes, particularly global warming, have altered the timing of harvest seasons in some countries. These changes may challenge vineyards by pushing them beyond climatic thresholds that make maintaining current wine quality more difficult. However, shifting weather patterns could also create opportunities for some regions to achieve more favourable conditions for producing well-balanced grapes.

The potential of any wine grape variety is fully realized only when the grapes are harvested at the optimal time, allowing the resulting wines to showcase their characteristic varietal aromas, flavours, and balance. Harvest timing is a critical decision influenced by management practices, environmental factors, and ongoing monitoring of fruit maturity. While the dates of previous harvests can serve as useful guidelines, they should not be solely relied upon due to the variability in growing conditions from season to season. Determining the right harvest time involves assessing several key parameters: sugar content, titratable acidity, pH, and other qualitative and quantitative measures. Proper evaluation of these factors ensures the production of high-quality grapes and wines.

The most commonly employed metric for determining harvest readiness is sugar content, measured in terms of total soluble solids (TSS), often expressed as degrees Brix (°Brix). Sugar concentration is crucial as it impacts both the sweetness of the fruit and the alcohol content of the wine. During fermentation, yeast converts these sugars into alcohol and carbon dioxide, with the alcohol level in the final wine directly correlating to the initial sugar levels in the grape juice. By monitoring and controlling the sugar concentration, winemakers can influence the balance and style of the wine, ensuring it meets the intended quality standards.

Titratable acidity (TA) measures the total quantity of acids present in the grape juice. Acidity is a vital component of a wine's structure, contributing to its freshness and balance. At harvest, the acidity levels must be sufficient to provide a firm backbone for the wine while complementing the sweetness and other flavour elements. Wines with well-balanced acidity are more vibrant and age better than those with imbalanced acid profiles.

The pH of grape juice is another critical parameter, affecting both the chemical stability and microbial health of the wine during fermentation and storage. A wine's pH impacts the solubility of tartrates, protein stability, and the rates of phenolic reactions that influence colour and mouthfeel. Additionally, pH influences the activity of microorganisms, making it a key factor in determining the wine's stability and longevity.

Phenolic compounds, including anthocyanins and tannins, are essential contributors to a wine's colour, structure, and flavour. These compounds play a role in browning reactions in grapes and wines and are critical to the wine's aging and maturation processes. Anthocyanins contribute to red wine's deep colour, while tannins provide structure, body, and astringency. Monitoring the development of phenolics during ripening ensures a desirable balance for the intended wine style.

In addition to sugar, acidity, and pH, other physical and sensory indicators are used to determine the right time for harvest. For example, the balance between sugar and acidity at harvest reflects the eventual ethanol and acidity balance post-fermentation. Soft berries with slightly dehydrated, softened pulp indicate ripeness. Ripe red grapes typically shift from green, herbaceous flavours reminiscent of asparagus or bell peppers to fruitier plum and cherry notes, signalling readiness for harvest.

Assessing the ripeness of grape pulp, skins, and seeds provides further insight into harvest timing. As grapes ripen, the pulp softens, and phenolic compounds in the skins transition from harsh and astringent to smooth and well-integrated. Skin tannins, which affect the wine's structure and body, reach an optimal balance during ripening, contributing to colour intensity and mouthfeel. Similarly, seeds transition from green to brown, becoming hard and developing nutty or toasty aromas, which are desirable traits in ripe grapes.

Another physical indicator of ripeness is the ease with which the pedicel, or grape stem, separates from the berry. Fully ripe grapes detach cleanly, leaving little to no pulp or skin tissue attached to the stem.

Harvest timing in viticulture is a multifaceted decision-making process that requires a thorough understanding of grape maturity. By evaluating sugar content, acidity, pH, and physical ripeness indicators such as berry texture, flavour, and phenolic development, growers can make informed decisions that optimize grape quality and ensure the production of exceptional wines. These assessments, combined with environmental monitoring and vineyard management practices, enable winemakers to craft wines that truly reflect the potential of their grape varieties and vineyard conditions.

When wine grapes are ready for harvest, winemakers face a crucial decision: whether to pick the grapes by hand or use mechanical harvesters. This decision is influenced by the vineyard's size, terrain, wine style, and the desired level of precision in fruit selection. Hand-picking remains the preferred method for many wineries, particularly for smaller vineyards and premium wine production, due to its precision, gentleness, and adaptability.

Hand-harvesting involves workers using secateurs to carefully clip grape bunches from the vines. The bunches are placed into small grape baskets or bins positioned along the rows beneath the vines. Workers ensure that the vines and the grapes remain undamaged during the process. Once the baskets are filled, the grapes are transported to larger containers or carefully loaded onto trucks for delivery to the winery. This method allows for meticulous handling, preserving the integrity of the fruit.

Figure 37: Hand harvesting Pinot noir grapes. Mark Smith, CC BY 2.0, via Wikimedia Commons.

One of the key advantages of hand-picking is the expertise and discernment of human workers, who can selectively harvest only the healthiest and ripest bunches. This level of precision is especially crucial for high-quality wine production, where even small imperfections can impact the final product. Hand-harvesting also minimizes damage to the grapes, which is particularly important for delicate varieties or styles of wine requiring intact berries.

Certain wine styles, such as dessert wines like Sauternes and Trockenbeerenauslese, necessitate hand-picking because individual berries must be carefully selected. These grapes, often affected by noble rot (botrytis), are fragile and prone to splitting if harvested mechanically. Hand-picking is also essential for producing late-harvest and ice wines, where selective harvesting ensures optimal ripeness and concentration.

The adaptability of hand-picking makes it suitable for any terrain, though it becomes more labour-intensive and costly on steep slopes, terraced hillsides, or uneven landscapes. Vineyards with Pergola-style trellising, common in northern Italy, or old bush vines with irregular growth patterns also require hand-harvesting. These scenarios make mechanization impractical or impossible.

A skilled picker can harvest up to two tonnes of grapes in a day, making hand-picking an efficient method despite its labour-intensive nature. It is particularly favoured in regions like New Zealand, where premium wine production emphasizes small, high-quality fruit from meticulously managed vineyards. In Europe, certain appellations, including Champagne, Beaujolais, Rioja, and Rive, have regulations mandating hand-harvesting to preserve traditional winemaking practices and ensure quality.

Moreover, some unique wine styles and techniques can only be achieved through hand-picking. For example, Beaujolais Nouveau requires intact grape bunches for carbonic maceration, while Amarone involves drying whole grape clusters to concentrate flavours. These methods demand careful handling of the fruit, which mechanical harvesting cannot provide.

Sustainable Harvesting Techniques

Sustainable grape harvesting practices are essential for minimizing environmental impact, maintaining vineyard health, and optimizing resource use while ensuring the production of high-quality grapes. Both manual and mechanical harvesting methods can be implemented sustainably, depending on the vineyard's size, terrain, and wine production goals.

Manual harvesting is characterized by its precision and selectivity, allowing workers to pick grapes at optimal ripeness and ensure only healthy fruit is collected. This method minimizes waste and maximizes harvest quality, which is particularly critical for premium wines. Studies indicate that manual harvesting can significantly reduce fossil fuel consumption and prevent soil compaction, thereby preserving soil structure and promoting long-term vineyard health [271, 366]. Furthermore, manual harvesting is adaptable to complex terrains, such as steep slopes and terraced landscapes, where machinery may not be feasible [146]. Sustainable hand-harvesting practices include the use of lightweight tools to reduce worker strain and biodegradable bins for grape transportation [367]. Additionally, a sustainable manual operation considers the well-being of workers by providing fair wages, safe working conditions, and ergonomic tools, ensuring a positive social impact while maintaining productivity [368]. Despite its labour-intensive nature and higher costs, training programs can enhance efficiency, and hybrid approaches that combine manual and mechanical methods can provide a balanced solution [369].

Figure 38: Grape harvesting machinery in operation at a vineyard in the Eden Valley, SA. CSIRO, CC BY 3.0, via Wikimedia Commons.

On the other hand, mechanical harvesting offers efficiency and speed, allowing for the rapid coverage of large vineyard areas and reducing the time grapes remain on the vine after ripening. This approach minimizes risks associated with over-ripening and can significantly lower labour costs [370]. Advances in technology have equipped modern harvesters with precision sensors that target specific grape clusters, reducing damage to vines and minimizing the collection of leaves or unripe fruit [371]. These innovations make mechanical harvesting increasingly suitable for premium wine production. Newer harvesters are designed for greater energy efficiency and may operate on alternative energy sources, thereby reducing carbon emissions [372]. Additionally, mechanical harvesting practices can mitigate soil compaction by utilizing lightweight designs or wider tires to evenly distribute weight, particularly when scheduled during dry conditions [270]. While mechanical harvesting is fast and cost-effective, it may lack the selectivity of manual methods. Hybrid systems, where workers sort grapes after mechanical harvesting, can address this limitation, ensuring high-quality outcomes while maintaining efficiency [274].

Both manual and mechanical methods can align with sustainable practices by timing the harvest to ensure grapes are picked at optimal ripeness, reducing waste, and enhancing wine quality. Waste reduction initiatives, such as composting grape stems and unmarketable fruit, contribute to natural fertilizers that enrich vineyard soil [373]. Water conservation practices, including the use of water-efficient cleaning systems, help minimize resource use [374]. Integrating renewable energy into tools and machinery further reduces the environmental footprint [271]. Additionally, sustainable harvesting involves maintaining biodiversity in and around the vineyard by protecting beneficial insects, using cover crops, and minimizing synthetic chemical use [368].

Sustainably managing a vineyard often requires a balanced combination of manual and mechanical harvesting. Large, flat areas may benefit from mechanical harvesters, while hand-picking is reserved for delicate grape varieties or challenging terrains. This hybrid approach ensures efficiency, preserves quality, and maintains environmental responsibility. By prioritizing soil health, reducing emissions, and fostering fair labour practices, both manual and mechanical harvesting methods contribute significantly to sustainable viticulture [375].

Managing Grape Waste: Composting and Reuse

The wine industry is characterized by the generation of substantial residuals and waste materials during the winemaking process, particularly during the harvest and production seasons. The primary by-products include grape pomace, lees, stalks, dewatered sludge, and winery wastewater. Grape pomace, which consists of seeds, stems, skins, and pulp left after juice extraction, constitutes approximately 20% of the solid waste generated in winemaking, while lees account for about 7% [376]. Annually, around 13 million tons of grape pomace are produced globally, underscoring the significant scale of this issue [376]. If not managed properly, these organic wastes can lead to environmental concerns such as soil and water pollution, particularly when disposed of in landfills [377].

Winery wastewater is another critical by-product, characterized by its organic matter content, low nitrogen and phosphorus levels, and an acidic pH ranging from 3 to 4. These properties complicate its management; however, the nutrient content presents opportunities for recycling and reuse [378]. The effective management of these residuals through innovative and eco-friendly strategies is essential for promoting sustainable waste management practices within the wine industry [377].

The suitability of grape pomace and other winery wastes for composting is well established. Grape pomace is rich in essential nutrients, including nitrogen, phosphorus, potassium, and calcium, making it an excellent soil conditioner [377]. Additionally, grape stalks and seed coats, which possess high carbon-to-nitrogen (C/N) ratios due to their lignocellulosic composition, contain tannins and polyphenols that can enhance compost quality [376]. Grape leaves and branches also contribute vital nutrients such as magnesium, sulphur, and potassium, further supporting their composting potential [377]. However, to optimize composting efficiency, pre-treatment of raw materials is necessary. For instance, shredding grape branches and leaves can accelerate decomposition, while the addition of nitrogen-rich materials can help balance the C/N ratio [377].

Figure 39: Close up of the pomace (grape skins, seeds, etc) left over after the wine has been pressed and juice removed. davitydave, CC BY 2.0, via Wikimedia Commons.

Windrow composting is a widely adopted method for managing grape and winery waste. This process involves arranging waste materials into long, narrow piles that are regularly turned to aerate the compost and maintain optimal conditions. Typically, the composting piles are about 1.5 meters high and 1.5 to 3 meters wide, with ideal moisture levels around 55%, maximum temperatures of 65°C, and oxygen concentrations between 5-10% [379]. The combination of pre-treated grape pomace, stalks, lees, and sludge in specific ratios helps achieve the desired C/N balance, facilitating microbial activity that decomposes the organic matter [379]. The composting process generally takes 7 to 15 days, resulting in a stable, dark brown material suitable for use as fertilizer [379].

The benefits of windrow composting extend beyond waste reduction; it can decrease the volume of organic waste by up to 40% and lower waste treatment costs [379]. The compost produced can be utilized as an organic fertilizer for grapevines and other crops, enhancing soil structure, increasing organic matter, and mitigating soil erosion [379]. By improving nutrient cycling, composting supports sustainable agricultural practices and reduces reliance on chemical fertilizers [379].

Chapter 8
Winemaking on a Small Scale

Basic Winemaking Techniques

Winemaking, or vinification, is an involved process that encompasses two broad categories: still wine production and sparkling wine production. The primary distinction between these categories lies in the presence or absence of carbonation. Still wines are produced without carbonation and represent the majority of global wine production, while sparkling wines contain carbonation, which can either be naturally occurring or artificially injected. This carbonation imparts effervescence, making sparkling wines distinctive and suitable for celebratory occasions.

The most common categories of wine are red, white, and rosé, which differ based on grape varieties, fermentation techniques, and contact with grape skins during production. Red wine is made from dark-skinned grapes and fermented with their skins to extract colour, tannins, and flavour compounds. White wine is produced from either light-skinned grapes or dark-skinned grapes with the skins removed before fermentation, resulting in a lighter, crisper profile. Rosé is crafted by limiting the contact time between the juice and grape skins, giving the wine a blush colour and a flavour profile that balances between red and white wines.

Although grapes are the primary raw material for winemaking due to their ideal balance of sugar, acidity, and tannins, wine can also be made from other plants. These wines, known as fruit wines, involve the fermentation of fruits like berries, cherries, or plums, offering a diverse range of flavours and styles.

Similar to wine but distinct in their raw materials and production methods are several light alcoholic beverages. Mead, for instance, is made by fermenting honey and water, sometimes with the addition of fruits, spices, or herbs. This ancient beverage is cherished for its unique sweetness and versatility.

Cider, often referred to as "apple cider," is made by fermenting the juice of apples. It ranges from dry to sweet and still to sparkling, depending on production techniques. Perry, or "pear cider," follows a similar process but uses pears as the primary ingredient, offering a lighter and fruitier alternative to cider.

Another notable fermented beverage is kumis, which is made by fermenting mare's milk. Though less common globally, kumis holds cultural significance in Central Asian countries, where it has been a traditional drink for centuries.

The winemaking journey begins with harvesting, where grapes are picked at optimal ripeness, determined by sugar, acid, and tannin levels. Once harvested, grapes are transported to the winery for primary fermentation preparation. The subsequent steps diverge depending on the type of wine being produced.

For red wine, fermentation involves the must—comprising grape juice, skins, seeds, and pulp. This process incorporates the grape skins, imparting colour, flavour, and tannins through maceration. Conversely, white wine is made by pressing the crushed grapes to separate the juice from the skins, which are discarded. In rare cases, white wine can be made from red grapes by minimizing skin contact. Rosé wines derive their pink hue from brief skin contact during fermentation or by blending red and white wines. Orange wine takes a unique approach, using white grapes but fermenting them with their skins, akin to red wine production.

Fermentation, the heart of winemaking, typically lasts one to two weeks. Yeast—either added or naturally occurring—converts grape sugars into alcohol and carbon dioxide. For red wines, this process occurs with the skins intact, while for white wines, it involves only the juice. After fermentation, red wines undergo a pressing stage to extract additional juice and wine from the skins. The resulting "free-run" wine may be blended with the press wine, depending on the winemaker's discretion.

For red wines, malolactic conversion is a bacterial process that softens the wine's taste by converting malic acid into lactic acid, reducing its sharpness. Many red wines are then aged in white oak barrels, where they acquire additional flavours and tannins. Before bottling, the wine is clarified, adjusted for balance, and prepared for maturation.

The timeline from harvest to consumption varies significantly. Some wines, like Beaujolais nouveau, are ready in a few months, while others, particularly those with high acid, tannin, or sugar, may improve over decades. However, only a minority of wines significantly benefit from extended aging.

Sparkling wines involve additional steps. In traditional methods, such as those used for Champagne, a secondary fermentation occurs within the bottle, trapping carbon dioxide and creating bubbles. The bottles are placed on riddling racks for sediment removal before being disgorged. Other methods, like the Charmat process, use sealed tanks for carbonation, and force-carbonation injects CO_2 directly into the wine, as commonly done with Prosecco.

Sweet wines retain residual sugar by halting fermentation before all sugars are converted to alcohol. Techniques include chilling the wine, adding sulphur, sterile filtration, or increasing initial sugar concentrations through methods like late harvesting, freezing grapes (ice wine), or using Botrytis cinerea fungus. Fortified wines like Port involve adding grape spirit to halt fermentation and preserve sugar, while the German technique of süssreserve involves reserving unfermented grape juice to sweeten the wine post-fermentation.

Winemaking generates by-products such as wastewater, pomace, and lees, requiring careful collection, treatment, and disposal or repurposing. These materials can be recycled as organic fertilizers, contributing to sustainability efforts in viticulture.

A modern innovation in winemaking is the production of synthetic wines, which bypass grapes entirely. These wines are engineered using water, ethanol, and additives like acids, amino acids, sugars, and organic compounds to replicate the taste and characteristics of traditional wines.

Winemaking is an art and science that encompasses diverse methods tailored to create various wine styles. Whether crafting traditional still wines, effervescent sparkling wines, or sweet and fortified varieties, winemakers employ a blend of ancient techniques and modern innovations. As the industry evolves, sustainable practices and emerging technologies like synthetic wines are shaping the future of this timeless craft.

Figure 40: Grapes sorting and crushing with winemaking equipment, Sokol Blosser Winery, Willamette Valley, Oregon. Howard Bales, CC BY 2.0, via Wikimedia Commons.

Crushing and primary fermentation are foundational steps in winemaking, crucial for extracting the juice and initiating the transformation of sugars into alcohol. These stages, whether done traditionally or with modern technology, play a pivotal role in defining the style and quality of wine.

Crushing and Destemming

Crushing involves gently breaking the skins of grapes to release their juices, while destemming removes the stems from the grape clusters. In ancient winemaking, grapes were crushed by foot, a practice still romanticized for its gentle handling. Today, small-scale operations may use manual or inexpensive crushers that can simultaneously destem. However, modern wineries often rely on mechanical crusher-destemmers to achieve efficiency and uniformity.

For white wines, the stems are typically retained during pressing to aid juice flow and minimize tannin extraction. Conversely, in red winemaking, stems are usually removed to prevent the release of harsh tannins and vegetal aromas unless the winemaker intentionally includes ripe stems to enhance tannic structure. Crushing may follow destemming to ensure optimal separation of skin and pulp without damaging the grape tissues excessively. For delicate varietals like Pinot Noir and Syrah, "whole berry" fermentation is sometimes employed to preserve fruity aromas through partial carbonic maceration.

Figure 41: Chardonnay grapes being pressed. Megan Mallen, CC BY 2.0, via Wikimedia Commons.

Primary Fermentation: Juice and Skins

The fermentation stage for red wines involves maceration, where the juice remains in contact with grape skins. This process extracts colour, flavour, and tannins, vital for red wine's structure and complexity. In white wine production, the juice is typically separated from the skins to avoid tannin extraction, preserving a clean and crisp profile. However, some winemakers opt for brief skin contact to enhance flavour, texture, or balance overly acidic juice.

Rosé wines are made by crushing red grapes and allowing short skin contact to achieve the desired colour before pressing and fermenting as white wine. Similarly, orange wines undergo extended maceration of white grapes, mirroring red wine techniques.

An alternative to traditional maceration is thermovinification, where grapes are heated to extract juice, tannins, and pigments. This method is particularly useful for red varietals and offers a pre-fermentation advantage in certain styles.

Yeast and Fermentation

Grape skins naturally harbor yeast, which can ferment the juice. While natural fermentation can add complexity, it may produce inconsistent results. To ensure controlled and reliable fermentation, many winemakers introduce cultured yeast. The primary fermentation lasts one to two weeks, during which yeast converts sugars into alcohol and carbon dioxide. The fermentation temperature significantly affects the wine's flavour profile, with reds fermenting at higher temperatures (22-25°C) for tannin extraction and whites at cooler temperatures (15-18°C) to retain delicate aromas.

Fermentation may also involve adjustments such as chaptalization, where sugar is added to boost alcohol content, or amelioration, which dilutes acidity by adding water and sugar. These techniques, subject to regional regulations, help achieve balanced wines in challenging growing conditions.

For red wines, a secondary bacterial fermentation, malolactic conversion, often follows primary fermentation. This process softens the wine by converting sharp malic acid into creamy lactic acid, enhancing mouthfeel and reducing acidity.

Figure 42: Pushing crushed Pinot noir must from the crusher into a fermentation tank where it will spend some time macerating on the skin, extracting colour, flavours, aromas and tannins during fermentation. Mark Smith, CC BY 2.0, via Wikimedia Commons.

Pressing separates liquid from solids, with red wines pressed after fermentation to extract any remaining juice from the skins. In white wine production, pressing occurs before fermentation to avoid excessive tannin extraction. Free-run juice, released without pressing, is of higher quality than press juice, which may contain more tannins and phenolics.

Modern presses, such as mechanical basket presses, offer precise control over pressure and extraction, preserving juice quality. Traditional manual presses, once widespread, are now primarily used for artisanal or high-end wines.

During red wine fermentation, grape skins and solids rise to form a "cap" on the liquid's surface. To ensure even extraction of tannins and colour, the cap must be regularly mixed into the juice. Pigeage, or cap punching, is a traditional French method where workers stomp the cap into the liquid. This process, now often mechanized, enhances the wine's structure and balance.

Primary fermentation and the handling of grapes and must shape the final wine's style and quality. From choosing skin contact duration to fermentation temperature and yeast selection, every decision influences the wine's flavour, texture, and structure. Whether using ancient techniques or modern innovations, winemakers tailor these processes to craft wines that express their vision and the unique characteristics of the grape.

Cold stabilization

Cold stabilization is a critical process in winemaking aimed at reducing the formation of tartrate crystals, which are also referred to as "wine crystals" or "wine diamonds." These crystals, composed of potassium bitartrate, occur naturally in wine as a result of the union of tartaric acid and potassium. While harmless, they may be perceived as undesirable by consumers when found in wine bottles, as they resemble sediment or grains of clear sand. Cold stabilization ensures a visually clear and aesthetically appealing product without altering the taste or quality of the wine.

Tartrate crystals form when tartaric acid, a naturally occurring compound in grapes, interacts with potassium, which is also present in the fruit. These compounds combine to create potassium bitartrate. The solubility of potassium bitartrate decreases at lower temperatures, causing it to precipitate out of the wine. While this precipitation is a natural occurrence, it can be problematic if it happens after the wine has been bottled, as it may confuse consumers or be mistaken for impurities.

Cold stabilization is typically performed after fermentation and before bottling. During this process, the wine is cooled to near-freezing temperatures, generally between -4°C to 0°C (25°F to 32°F), for a period of 1–2 weeks. This chilling encourages the potassium bitartrate to crystallize and separate from the wine. The crystals often adhere to the walls of the storage tank or vessel used during the process.

Once the desired stabilization period is complete, the wine is carefully racked or drained, leaving the crystals behind. The removal of these crystals minimizes the risk of their formation in the final bottled product, particularly when the wine is exposed to cold storage conditions during distribution or consumption.

The primary advantage of cold stabilization is to improve the visual clarity and marketability of wine. By eliminating the potential for tartrate crystals to form in the bottle, winemakers ensure a product that meets consumer expectations for clarity and quality. This process is particularly important for white and rosé wines, where clarity is highly valued, though it is also performed on certain red wines.

Cold stabilization also enhances the wine's stability during storage and transportation. Wines that undergo this process are less likely to form tartrate crystals even if exposed to varying temperatures, ensuring consistency in the product.

While cold stabilization is effective, it requires significant energy consumption to maintain the necessary low temperatures, which can impact production costs and environmental sustainability. As a result, some wineries are exploring alternative methods, such as using additives like carboxymethyl cellulose (CMC) or ion exchange resins, to achieve similar stabilization results with reduced energy usage.

Figure 43: The wines in this picture are going through cold stabilization with the temperature being brought down to at least -4C (25F) to bring the tartrate crystals out of solution and remove them from the wine. Photo taken at Columbia Crest Winery in Patterson, Washington. Agne27, CC BY-SA 3.0, via Wikimedia Commons.

Moreover, the process must be carefully managed to avoid over-stabilization, which can strip the wine of its natural acidity and alter its balance. Winemakers must monitor the wine closely to ensure the desired outcome without compromising its flavour profile.

Despite cold stabilization, tartrate crystals can still form in bottled wine if it is exposed to extremely cold temperatures, such as in refrigerators or cold storage facilities. While these crystals are harmless and do not affect the taste or quality of the wine, they may still be perceived as a flaw by uninformed consumers. Educating consumers about the natural occurrence of these crystals can help mitigate negative perceptions.

Secondary (Malolactic) Fermentation and Bulk Aging

Secondary fermentation and bulk aging are pivotal stages in the winemaking process that refine the wine's flavour, texture, and stability. These stages typically last between three to six months, during which the wine undergoes gradual changes under carefully controlled conditions. The primary objective is to enhance clarity, balance acidity, and develop complex flavours.

Secondary fermentation, often referred to as malolactic fermentation (MLF), is a biological process where lactic acid bacteria convert the sharper malic acid in the wine (reminiscent of green apples) into softer lactic acid (reminiscent of cream). This process not only softens the wine's acidity but also adds creamy, buttery notes and enhances its mouthfeel. While MLF is common in red wines, it is selectively applied to white wines like Chardonnay to achieve specific stylistic goals. Wines undergoing MLF are kept under an airlock to prevent oxygen exposure, which could lead to spoilage.

During this phase, proteins from the grapes are broken down, and residual yeast cells along with other grape particles settle at the bottom of the vessel, forming a layer known as lees. Potassium bitartrate also begins to precipitate, which can be further managed through cold stabilization to prevent tartrate crystal formation in the final product. As these sediments settle, the wine transitions from cloudy to clear.

To improve clarity and remove lees, the wine is periodically racked—transferred from one vessel to another—leaving behind the sediment. This process reduces the risk of off-flavours developing from autolysis (the breakdown of dead yeast cells) while contributing to the wine's brightness and finesse. Racking also allows winemakers to aerate the wine minimally, which can help soften tannins in red wines.

After secondary fermentation, the wine is bulk-aged to allow its flavours and aromas to integrate fully. Aging typically occurs in large vessels such as stainless steel tanks, oak barrels, or glass demijohns, with the choice of vessel significantly influencing the wine's final profile.

- **Stainless Steel Tanks**: These neutral vessels preserve the wine's freshness and primary fruit characteristics, making them ideal for unoaked wines.

Stainless steel also allows precise temperature control, which is crucial for preventing spoilage and promoting slow, steady maturation.

- **Oak Barrels**: Aging in oak barrels adds layers of complexity to the wine. Oak imparts flavours such as vanilla, spice, and toast, and can contribute tannins that enhance structure and aging potential. Winemakers may use new oak barrels for stronger flavour influence or older barrels for subtle integration.

- **Glass Demijohns**: Often used by amateur winemakers or for small batches, glass demijohns allow clear visibility of the wine's development. They are inert, ensuring no additional flavours are imparted, making them suitable for preserving the purity of the wine's natural characteristics.

In some cases, oak alternatives like chips or staves are used with non-wooden barrels to mimic the effects of traditional barrel aging, particularly in cost-effective wine production.

The choice of aging vessel depends on the type of wine being produced, the grape variety, and the winemaker's stylistic preferences. For example, unoaked wines intended to highlight fresh, fruity notes are typically aged in stainless steel, while wines aimed at achieving richness and complexity may spend significant time in oak barrels. Winemakers may also blend wines aged in different vessels to achieve a balance of freshness and complexity.

By the end of secondary fermentation and bulk aging, the wine undergoes significant transformation. Its flavours become more integrated, acidity is balanced, and its texture is refined. The wine emerges clearer, more stable, and ready for bottling or further aging, depending on the winemaker's goals. This stage is critical for producing a high-quality wine that meets the desired flavour profile and aging potential.

Malolactic fermentation

Malolactic fermentation (MLF) is a critical process in winemaking where lactic acid bacteria convert malic acid into lactic acid and carbon dioxide. This transformation not only alters the chemical composition of the wine but also significantly impacts its taste, texture, and stability. MLF can be deliberately induced by introducing cultured strains of lactic acid bacteria or may occur spontaneously if natural lactic acid bacteria are present in the wine.

Malic acid, naturally present in grapes, is a stronger acid with two carboxylic acid groups (-COOH) that contribute to a wine's sharp and tart taste. During MLF, this malic acid is metabolized into lactic acid, a gentler acid with only one carboxylic acid group. This conversion reduces the overall acidity of the wine and softens its flavour profile, creating a smoother and rounder mouthfeel. Wines with high concentrations of malic acid, which can taste harsh and bitter, benefit greatly from this process. The lactic acid produced during MLF is commonly associated with dairy products, contributing to the creamy texture often found in certain wines.

The process also lowers the wine's total acidity, which can be adjusted if needed by adding tartaric acid to maintain a balanced pH. Monitoring the pH is essential during MLF, as excessively high pH levels can lead to microbial instability. Ideal pH levels are typically below 3.55 for white wines and 3.80 for red wines.

Malolactic fermentation plays a significant role in shaping the flavour profile of wine. One of the by-products of MLF is diacetyl, a compound that imparts a buttery aroma and taste. This is particularly noticeable in certain styles of Chardonnay, where winemakers intentionally emphasize the buttery character by managing diacetyl levels. The extent of diacetyl production can be controlled by adjusting the bacteria strain, fermentation temperature, and timing of sulphur dioxide additions.

In red wines, MLF enhances the wine's complexity and softness by reducing sharp acidity and contributing subtle flavour notes. For white wines, the decision to employ MLF depends on the desired style. Aromatic whites like Riesling and Sauvignon Blanc often avoid MLF to preserve their crispness and bright acidity. Conversely, fuller-bodied whites like barrel-fermented Chardonnay frequently undergo MLF to achieve a richer, creamier texture.

Most red wines are subjected to complete malolactic fermentation for both sensory and stability reasons. Allowing MLF to occur ensures that any remaining malic acid does not undergo spontaneous fermentation later, such as in the bottle, which could create unwanted carbonation or off-flavours. White wines, however, may undergo MLF selectively based on stylistic preferences. For example, winemakers might employ partial malolactic fermentation, fermenting less than 50% of the wine, to balance acidity with softness while retaining some of the wine's original brightness.

MLF typically occurs after primary fermentation, during the aging phase of winemaking. It is conducted under controlled conditions to ensure the optimal activity of the lactic acid bacteria. Factors such as temperature, sulphur dioxide levels, and pH are carefully managed to create an environment conducive to MLF while preventing spoilage.

Malolactic fermentation is a transformative step in winemaking that enhances the sensory qualities of wine while ensuring its stability. By converting tart malic acid into softer lactic acid, MLF reduces acidity, adds complexity, and shapes the wine's texture. While it is a standard practice for most red wines, its application in white wines varies depending on the style. Whether complete, partial, or avoided altogether, MLF allows winemakers to craft wines with diverse flavour profiles and balanced acidity, meeting the preferences of a wide range of consumers.

Laboratory Testing

Wine aging, whether in tanks or barrels, involves continuous monitoring and testing to ensure the final product meets the winemaker's standards for quality, stability, and taste. These laboratory tests provide vital data, allowing winemakers to make adjustments throughout the winemaking process. The tests cover a range of chemical, physical, and sensory parameters, offering insights into the wine's development and its readiness for bottling.

Key tests performed during wine aging:

Brix (°Bx) and Sugar Content: Brix is a measure of the soluble solids in grape juice or wine, primarily indicating sugar levels, though it includes other dissolved substances like salts, acids, and tannins. Measured in grams per 100 grams of solution, it directly relates to the potential alcohol content and grape maturity. For example, 20 °Bx means 20 grams of dissolved compounds in 100 grams of juice. Refractometers and hydrometers are commonly used for measuring Brix, with hydrometers offering a cost-effective alternative. The French Baumé scale (Bé°) is another method used, where one Bé° corresponds approximately to one percent alcohol. Chaptalization, the practice of adding sugar to achieve a desired alcohol level, is regulated or prohibited in some regions, emphasizing the importance of accurate sugar measurement in legal and quality compliance.

Volatile Acidity (VA): Volatile acidity measures steam-distillable acids, primarily acetic acid, which is the main component of vinegar. Excessive VA can indicate microbial spoilage and negatively affect wine quality. While small amounts of acetic acid occur naturally, its presence in higher concentrations usually results from microbial metabolism, particularly from acetic acid bacteria. Tests for VA can be performed using techniques like cash still distillation, HPLC, gas chromatography, or enzymatic methods. Preventative measures include eliminating oxygen from wine containers, using sulphur dioxide (SO_2) to inhibit bacterial growth, and rejecting mouldy grapes. In cases where VA is excessive, remedies include reverse osmosis, blending with low-VA wine, or filtering to remove the responsible microbes.

Sulphur Dioxide (SO_2) Levels: Sulphur dioxide is crucial in winemaking for its antioxidant and antimicrobial properties. Winemakers test for free or available SO_2 and total SO_2 to ensure optimal levels for wine preservation and to avoid off-flavours or spoilage. A typical test involves acidifying a wine sample, distilling the liberated SO_2, and capturing it in a hydrogen peroxide solution. The resultant sulfuric acid is then titrated with sodium hydroxide (NaOH) to determine the SO_2 concentration. This method, while reliable, has challenges in red wines and with inefficient equipment. Despite minor inaccuracies, the test offers sufficient precision (2.5–5% error) to maintain SO_2 levels within desired ranges.

pH and Titratable Acidity (TA): pH measures the hydrogen ion concentration, indicating the wine's acidity level. Lower pH values suggest higher acidity, which is essential for wine stability and freshness. Titratable acidity quantifies the total acid content, directly influencing taste and structure. Monitoring pH and TA allows winemakers to adjust acidity using tartaric acid or blending techniques, ensuring balance in the final product.

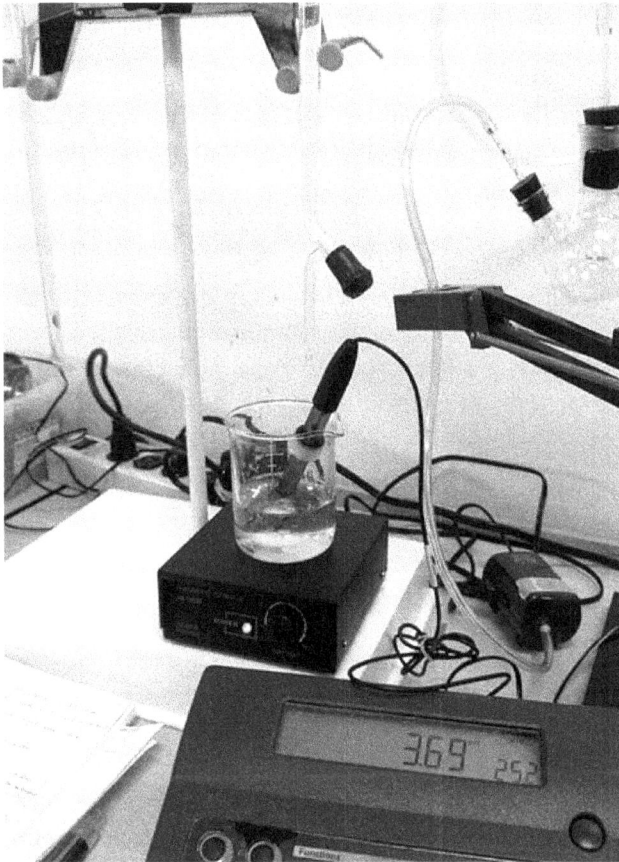

Figure 44: Winery lab bench work during a Titration Assay (TA) test to determine the acidity of a wine. The wine sample is diluted with distilled water and titrated with NaOH while the pH is measured with a pH meter. Agne27, CC BY-SA 3.0, via Wikimedia Commons.

Additional tests are performed for specific quality and stability parameters. These include:

- **Residual Sugar**: Determines sweetness levels, which influence fermentation completeness and style.

- **Alcohol Content**: Impacts the wine's body, flavour profile, and regulatory compliance.

- **Potassium Hydrogen Tartrate (Cream of Tartar)**: Tests for the potential crystallization of tartrates, often mitigated by cold stabilization.

- **Heat-Unstable Proteins**: Conducted primarily in white wines to prevent haze formation after bottling.

In addition to chemical tests, sensory evaluations are conducted throughout aging. Winemakers assess the wine's aroma, taste, and mouthfeel, allowing them to identify potential issues or areas for improvement. For instance, if a wine exhibits a harsh taste, proteins or other fining agents may be added to soften the texture. Sensory tests are essential for aligning the wine with the desired flavour profile and style.

The data obtained from these tests guide winemaking decisions. For example, if sulphur dioxide levels are low, additional SO_2 can be added to protect the wine. High volatile acidity may require blending or microbial intervention. Monitoring Brix and residual sugar ensures proper fermentation and alcohol levels, while adjustments to pH and TA maintain balance and stability. These proactive measures allow winemakers to refine the wine's quality and avoid faults before bottling.

Blending and Fining

Blending and fining are two essential processes in winemaking that significantly impact the final product's taste, clarity, and stability. These techniques allow winemakers to fine-tune their wines, ensuring they meet desired quality standards and market expectations.

Blending involves mixing different batches of wine to achieve a specific taste, texture, and aroma profile. It is a highly creative and technical process that provides winemakers with the flexibility to correct or enhance the wine's characteristics.

One common reason for blending is to balance the wine's acid and tannin levels, which affect its structure and mouthfeel. For example, if a particular batch is overly acidic, blending it with a wine with lower acidity can create a more balanced flavour. Similarly, a wine with a lacklustre body can be enhanced by blending it with a fuller-bodied wine.

Blending also allows winemakers to create consistency across vintages. By mixing wines from different grape varieties, regions, or harvest years, they can ensure that their product maintains a recognizable flavour profile that aligns with consumer expectations. For example, Champagne production frequently employs blending from multiple vintages to produce non-vintage wines with consistent taste year after year.

Moreover, blending can be used to introduce complexity to the wine. By combining different grape varieties, such as Cabernet Sauvignon with Merlot in a Bordeaux blend, winemakers can layer diverse flavours, aromas, and textures, creating a multidimensional wine. The timing of blending—whether before or after fermentation or aging—also influences the wine's final profile.

Figure 45: Bench trail set up to test the dosage of isinglass and bentonite fining agent needed for winemaking. Agne27, CC BY-SA 3.0, via Wikimedia Commons.

Fining is the process of clarifying and stabilizing wine by removing unwanted substances such as tannins, proteins, and microscopic particles that can cause cloudiness or affect taste. Fining agents are introduced into the wine to bind with these impurities, forming larger particles that can be easily removed through filtration or settling.

Gelatin is one of the most widely used fining agents and has been employed in winemaking for centuries. It is particularly effective at reducing tannin content and softening astringency, especially in red wines. When added to the wine, gelatin reacts with tannins and other wine components to form a sediment, which is then removed through filtration.

Other animal-derived fining agents include casein (a milk protein), egg whites, and isinglass (derived from fish bladder). These agents are valued for their ability to clarify wine while preserving its delicate flavours. For instance, egg whites are often used in high-quality red wines to gently refine tannins without stripping away desirable characteristics.

To cater to vegan and vegetarian consumers, non-animal-based fining agents have become increasingly popular. Bentonite, a volcanic clay, is a common choice, particularly in white wines, as it effectively removes proteins that can cause haze. Other non-animal agents include diatomaceous earth, cellulose pads, and membrane filters. These options provide winemakers with effective alternatives while ensuring the wine remains suitable for all consumers.

Fining agents serve multiple purposes beyond simply clarifying the wine. They help stabilize the wine, preventing unwanted changes during storage and transportation. For instance, proteins in wine can precipitate over time, causing haziness. Fining with bentonite removes these proteins, ensuring the wine remains visually appealing and shelf-stable.

The choice of fining agent depends on the specific needs of the wine. For example, red wines with high tannin levels may benefit from gelatin or egg white fining, while white wines may require bentonite to address protein stability. Winemakers carefully tailor their approach based on the wine's composition and the desired final product.

Some wines, such as aromatized wines, incorporate additional ingredients like honey, egg-yolk extract, or woody extracts to achieve specific flavours or aging characteristics. These practices add unique nuances to the wine but require precise control to maintain balance and quality.

Winemakers must also consider the preferences and expectations of their target audience. For instance, using non-animal fining agents aligns with the growing demand for vegan-friendly wines, while traditional methods may be preferred for wines targeting connoisseurs who value historical authenticity.

Preservatives

Preservatives play a vital role in winemaking by protecting the wine from spoilage, oxidation, and microbial activity. The most commonly used preservative is sulphur dioxide (SO_2), added in various forms such as liquid sulphur dioxide, sodium metabisulfite, or potassium metabisulfite. Another preservative, potassium sorbate, is occasionally used for specific purposes, especially in sweet wines. These substances ensure that the wine retains its quality and remains stable throughout its storage and transportation.

Sulphur dioxide serves two primary purposes in winemaking. First, it acts as an antimicrobial agent, inhibiting the growth of bacteria and undesirable yeast that could spoil the wine. Second, it functions as an antioxidant, preventing oxidation that could degrade the wine's flavour, colour, and aroma.

In white winemaking, sulphur dioxide is typically added before fermentation and immediately after alcoholic fermentation. When added after fermentation, it helps to prevent malolactic fermentation (MLF), a secondary fermentation that can alter the wine's acidity and flavour profile. It also protects against bacterial spoilage and oxygen exposure. For white wines, the free sulphur dioxide level is maintained at around 30 mg/L until bottling. For rosé wines, even smaller additions are preferred, as excessive SO_2 can interfere with the delicate flavour and colour balance. The goal is to ensure the available SO_2 does not exceed 30 mg/L.

In red winemaking, sulphur dioxide may be used at higher levels (up to 100 mg/L) before fermentation to stabilize colour by helping extract anthocyanins from grape skins. After malolactic fermentation, it performs similar functions as in white winemaking, protecting against spoilage and oxidation. However, red wines require lower maintenance levels of SO_2, usually around 20 mg/L, to avoid bleaching pigments and altering the wine's natural colour. Small additions can also be made after alcoholic fermentation but before MLF to manage minor oxidation and prevent bacterial growth.

Maintaining appropriate SO_2 levels is crucial. The free sulphur dioxide concentration should be carefully measured using methods such as aspiration. Consistently high levels may cause sensory issues, such as a sharp, unpleasant aroma, while insufficient levels leave the wine vulnerable to spoilage. Winemakers adjust SO_2 based on the wine's type, pH, and storage conditions.

Potassium sorbate is primarily used in sweet wines to control fungal growth, particularly yeast, which can reinitiate fermentation in the bottle. This makes it an effective tool for stabilizing wines with residual sugar.

One downside of potassium sorbate is the risk of its conversion into geraniol, a compound with a strong, unpleasant aroma. This occurs if malolactic fermentation takes place in the presence of sorbate. To prevent this, winemakers either ensure the wine undergoes sterile filtration before bottling or maintain sufficient sulphur dioxide levels to inhibit bacterial activity.

Sterile bottling involves filtering the wine to remove all microbial life, ensuring long-term stability without the risk of unwanted fermentation. This method is particularly useful for wines treated with potassium sorbate, as it prevents the conditions that lead to geraniol formation.

Some winemakers opt for natural winemaking techniques that exclude preservatives such as sulphur dioxide or potassium sorbate. Instead, these wines rely on rigorous hygiene practices, low-temperature storage, and careful monitoring. Once bottled, natural wines are often refrigerated at around 5°C (41°F) to slow down microbial activity and oxidation. While this approach appeals to consumers seeking minimal intervention wines, it carries a higher risk of spoilage and requires meticulous handling.

Without the use of preservatives like sulphur dioxide, wines are highly susceptible to bacterial spoilage, regardless of how hygienic the winemaking process is. This is particularly true for wines with higher pH levels, which provide a more favourable environment for microbial growth. Preservatives ensure that wines remain stable, age gracefully, and reach consumers in optimal condition.

Filtration

Filtration plays a pivotal role in winemaking, primarily serving two essential functions: clarification and microbial stabilization. These processes are crucial for enhancing the wine's visual appeal, stability, and overall quality.

The clarification process aims to eliminate larger particles that can cloud the wine, thereby affecting its visual clarity. Particles such as grape skins, seeds, and yeast cells are common culprits that necessitate filtration. The clarification process can be divided into several stages. Initially, coarse polishing removes particles larger than 5–10 micrometres, which is essential for preparing the wine for finer filtration [380]. Subsequently, a more refined filtration step targets particles larger than 1–4 micrometres, ensuring a clean and polished appearance [381].

Natural clarification methods, such as cold stabilization, are also employed, where wine is refrigerated at approximately 35°F (2°C) for a month. This allows particles to settle naturally due to gravity, requiring no chemical intervention, and is particularly favoured for wines that undergo minimal processing [382]. The use of pectinases can further facilitate this process by degrading structural polysaccharides that interfere with filtration, thereby enhancing the release of colour and aroma compounds [383].

Microbial stabilization is another critical aspect of filtration, focusing on reducing yeast and bacteria levels to prevent spoilage and re-fermentation. While complete sterility is not the goal, effective filtration can significantly lower microbial loads to ensure wine stability during storage and transport. For yeast retention, a pore size of at least 0.65 micrometres is necessary, while a finer pore size of 0.45 micrometres is required for bacteria [384]. This filtration process, however, can affect the wine's colour and body, particularly in red wines, necessitating a balance between stabilization and the preservation of sensory characteristics [385].

Filtration for microbial stabilization should complement other winemaking practices, such as maintaining proper hygiene and using sulphur dioxide, rather than replace them. This multifaceted approach ensures that the wine remains stable while retaining its natural complexity and body, especially in premium or artisanal wines [386].

The choice of filtration method and its extent depend significantly on the winemaker's objectives and the desired wine style. For instance, heavy filtration may be appropriate for commercial wines requiring long-term stability, while minimal filtration might be preferred for premium wines to preserve their natural characteristics [387]. The interplay of various factors, including wine composition and the type of filtration media used, can influence the efficiency and effectiveness of the filtration process [388].

Moreover, innovative filtration techniques, such as cross-flow microfiltration, have been developed to enhance efficiency while minimizing the impact on the wine's sensory attributes [388]. These methods can help manage fouling, which occurs due to the deposition of wine constituents on the filter, thereby affecting the filtration performance [388].

Bottling

Bottling marks the culmination of the winemaking process, where the wine is sealed and prepared for distribution and consumption. This step is crucial not only for preserving the wine's quality but also for ensuring its safety during storage and transport. The process involves several steps, from adding preservatives to sealing and labelling the bottles.

Before bottling, a final dose of sulphite is typically added to the wine. Sulphites act as preservatives, safeguarding the wine against oxidation and microbial activity that could lead to spoilage or unwanted fermentation inside the bottle. The precise amount of sulphite is carefully measured to strike a balance between preservation and maintaining the wine's natural flavours. Overuse of sulphites can negatively affect the taste, while insufficient amounts may compromise the wine's longevity.

Traditionally, wine bottles are sealed with natural corks, a practice that has been in use for centuries. Corks are prized for their ability to allow small amounts of oxygen into the bottle, which aids in the aging process for certain wines. However, natural corks are prone to cork taint, a flaw caused by the presence of compounds like TCA (trichloroanisole), which can ruin the wine's aroma and flavour.

To mitigate this issue, alternative closures have gained popularity:

- **Synthetic Corks**: Made from plastic or other synthetic materials, these are less expensive than natural corks and eliminate the risk of cork taint. However, they may not offer the same aging potential as natural corks.

- **Screw Caps**: Increasingly popular for their reliability and convenience, screw caps create a tight seal and prevent oxidation. While once considered a marker of lower-quality wine, they are now widely accepted for preserving the freshness of both young and aged wines.

Each closure method has its advantages and is selected based on the wine type, intended storage duration, and target market.

After sealing the bottle, a capsule is applied to the top for aesthetic and functional purposes. Capsules, which are typically made of foil, aluminium, or plastic, provide an additional layer of protection against contamination and improve the bottle's visual appeal. Once placed, the capsule is heated to shrink tightly over the neck of the bottle, creating a secure and polished finish. This step also helps to reinforce the seal and prevent tampering.

In addition to traditional glass bottles, some wines are packaged in bags or boxes, often referred to as "bag-in-box" wines. These alternatives are especially popular for their convenience, cost-effectiveness, and reduced environmental footprint. Bag-in-box packaging involves placing the wine in a sealed plastic bag, housed within a cardboard box. This method minimizes air exposure, preserving the wine's freshness even after opening.

While some consumers may perceive boxed wines as lower quality, advancements in packaging technology have made it a viable option for a range of wines, including premium selections. The choice of packaging often depends on the wine's intended use, with boxed wines commonly used for casual consumption and events.

The choice of bottling and closure methods also reflects the winemaker's goals regarding aging potential and market appeal. For wines meant to age for several years, closures that allow gradual oxygen exchange, such as natural corks, are preferred. For wines intended for immediate consumption, screw caps or synthetic corks are often used to preserve freshness.

Figure 46: Chenin blanc wine from Washington State being filled in the bottling truck. Agne27, CC BY-SA 3.0, via Wikimedia Commons.

In addition to functionality, the design of the bottle and closure contributes to the wine's branding and market positioning. A well-sealed, attractively packaged bottle not only preserves the wine but also enhances its appeal to consumers.

Once bottled, wines undergo final quality checks to ensure consistency and adherence to regulatory standards. Bottled wines are then labelled with all required information, including varietal, vintage, region of origin, and alcohol content, before being distributed to retailers or directly to consumers.

Bottling is not merely a logistical step; it is a critical phase where preservation, presentation, and practicality converge to deliver a product that reflects the winemaker's dedication and expertise. Whether sealed with corks, screw caps, or alternative closures, or packaged in bottles, bags, or boxes, the ultimate goal is to protect the wine's integrity and enhance the consumer's experience.

Organic and Natural Winemaking Practices

Organic and natural winemaking are approaches that emphasize sustainability, environmental stewardship, and minimal intervention in both the vineyard and winery. While both methods aim to produce wine with as few additives and processes as possible, they differ in their guidelines, certifications, and philosophies.

Organic and natural winemaking are both rooted in the principles of sustainability and environmental responsibility, but they differ significantly in their practices and approaches. One key distinction lies in regulation. Organic winemaking is governed by certified standards that dictate vineyard and winery practices, including the use of additives, farming methods, and labelling requirements. In contrast, natural winemaking operates as a philosophical approach with no formal regulations or universally recognized certifications. This makes natural winemaking more open to interpretation and less standardized.

Another major difference is the use of additives. Organic winemaking permits the use of certain additives, such as limited quantities of sulphur dioxide, to stabilize and preserve the wine. These additions are carefully controlled under organic certification guidelines. Natural winemaking, on the other hand, avoids additives entirely, relying instead on the natural composition of the grapes and spontaneous fermentation to guide the process. This lack of intervention often leads to a more unpredictable and unique wine profile.

The approaches also diverge in processing techniques. Organic wines may undergo processes like filtration and fining to clarify the wine and remove impurities, provided these practices align with organic certification standards. In contrast, natural winemaking typically avoids such interventions altogether. Natural wines are often bottled unfiltered and unfined, resulting in a cloudy appearance and sediment, which some consumers view as a mark of authenticity and minimal interference. These differences reflect the broader philosophies of each approach, with organic winemaking focusing on regulated sustainability and natural winemaking emphasizing purity and minimalism.

Organic Winemaking Practices

Organic winemaking practices are characterized by a commitment to sustainable agricultural methods that prioritize environmental health, biodiversity, and the avoidance of synthetic chemicals throughout vineyard management, grape production, and winemaking processes. This synthesis will explore the key components of organic winemaking, including vineyard management, grape production, winemaking practices, and certification.

In organic viticulture, soil health is paramount. Organic vineyards eschew synthetic fertilizers, opting instead for natural amendments such as compost, green manure, and cover crops. These practices enrich the soil with essential nutrients and enhance its structure, promoting a healthier ecosystem [284]. Furthermore, pest and disease control in organic vineyards relies on natural solutions. The use of synthetic pesticides and herbicides is strictly prohibited; instead, organic vineyards may introduce beneficial insects, apply organic sprays like sulphur and copper, and implement crop rotation strategies to manage pests and diseases effectively [389]. Additionally, promoting biodiversity is a critical aspect of organic vineyard management. By planting hedgerows and wildflowers, organic vineyards create habitats for beneficial organisms, which can help reduce pest populations naturally [98].

Organic certification mandates that no genetically modified organisms (GMOs) are used in grape cultivation, ensuring that the grapes are grown in a manner that aligns with organic principles [390]. Harvesting practices in organic vineyards typically involve hand-picking, which not only ensures the quality of the grapes but also minimizes environmental impact by reducing machinery use [391]. Research has shown that grapes produced under organic conditions often exhibit higher levels of bioactive compounds, which can enhance their nutritional quality compared to conventionally grown grapes [392, 393].

In the winemaking process, organic regulations impose strict limits on the use of additives. While sulphur dioxide (SO_2) is permitted to prevent oxidation and spoilage, its use is restricted to much lower levels than in conventional winemaking [391]. Additionally, processes such as filtration and fining must utilize organic or non-animal-based materials if the wine is labelled as "vegan organic" [391]. The emphasis on minimal intervention during winemaking aligns with the overall philosophy of organic practices, which seeks to preserve the natural characteristics of the grapes and the terroir [394].

To label wine as organic, producers must comply with the standards set by recognized organic certifying bodies, such as the USDA Organic in the United States or EU Organic in Europe. These standards encompass all aspects of production, from vineyard management to bottling, ensuring that the entire process adheres to organic principles [395]. The certification process is crucial for maintaining consumer trust and ensuring that organic wines meet the expected environmental and health standards [395].

Natural Winemaking Practices

Natural winemaking represents a holistic approach to viticulture and enology that transcends organic principles, emphasizing minimal intervention throughout the entire winemaking process. This philosophy is not merely a set of practices but rather a conceptual framework that prioritizes the integrity of the wine's natural characteristics. Natural winemakers seek to encapsulate not only the terroir—defined by climate, soil, and vine characteristics—but also their ethical commitments within the wine itself, thereby creating a product that reflects both environmental and cultural values [396]. This perspective is supported by the observation that the natural wine movement lacks universally recognized certifications, leading to a diverse range of practices and interpretations among winemakers [397].

In terms of vineyard management, most natural winemakers start with organic or biodynamic practices, which focus on enhancing soil health, promoting biodiversity, and employing natural pest control methods. This foundation is crucial as it aligns with the principles of sustainability that many natural winemakers advocate, which include reducing water usage, conserving energy, and minimizing the carbon footprint of their operations [398]. The emphasis on organic and biodynamic practices is not only about compliance but also about fostering a deeper connection with the land, which is essential for producing wines that authentically represent their origins (Padilla et al., 2016).

When it comes to the winemaking process itself, natural wines are characterized by the exclusion of additives, including sulphur dioxide, although some winemakers may use minimal sulphur at bottling to stabilize the wine [398]. The fermentation process relies exclusively on native yeasts, which are naturally present on grape skins or in the winery environment, rather than on cultured yeasts. This reliance on indigenous yeasts is critical for preserving the unique microbial terroir of the vineyard, which contributes to the wine's distinctive flavour profile [399, 400]. Furthermore, natural wines are typically not filtered or fined, resulting in a cloudy appearance and the potential for sediment, which are often seen as indicators of authenticity and minimal intervention [397].

The characteristics of natural wines can be quite unpredictable, often showcasing unconventional flavours that arise from the lack of intervention during the fermentation process. This unpredictability can lead to a shorter shelf life, as the absence of preservatives makes these wines more susceptible to spoilage [397]. Additionally, the cloudy appearance of natural wines, a result of minimal filtration, is often embraced by consumers as a sign of authenticity and a connection to traditional winemaking practices [398].

Challenges of Organic and Natural Winemaking

Organic and natural winemaking, while celebrated for their environmental and philosophical principles, present several challenges that require thoughtful solutions. These challenges impact vineyard management, production processes, and marketability, posing hurdles that winemakers must navigate to produce high-quality, sustainable wines.

One of the most significant challenges in organic and natural winemaking is the increased vulnerability of vineyards to pests, diseases, and adverse climate conditions. Without the use of synthetic chemicals, organic and natural vineyards rely on natural pest control methods and disease management strategies, which may not always be as immediately effective. This can result in crop losses or reduced grape quality.

To overcome these vulnerabilities, winemakers can adopt integrated pest management (IPM) techniques, which combine biological controls, crop rotation, and habitat creation for beneficial insects. The use of disease-resistant grape varieties and the implementation of precision agriculture tools, such as drones and sensors, can help monitor and address issues before they escalate. Additionally, practices like cover cropping and composting enhance soil health, making vines more resilient to environmental stressors.

Natural wines often exhibit significant variability between vintages due to the minimal intervention approach. This lack of consistency can deter consumers who value predictability in wine quality. Factors such as wild fermentation and the absence of additives like sulphur dioxide contribute to this variability, making natural wines more susceptible to spoilage and off-flavours.

Winemakers can address this challenge by improving their understanding of vineyard-specific factors, such as soil composition and microclimate, and how they influence grape characteristics. While maintaining the natural philosophy, producers can use controlled environments for fermentation and invest in advanced monitoring technologies to track and manage microbial activity. Offering transparent communication about the natural winemaking process can also help set consumer expectations and build appreciation for the wine's unique character.

For organic winemakers, obtaining certification can be a costly and time-consuming process, especially for small producers. Certification involves adhering to stringent guidelines, extensive record-keeping, and inspections, which can strain financial and operational resources. The challenge is compounded for producers in regions where organic practices are not widely supported or subsidized.

To mitigate these costs, small-scale producers can collaborate with cooperative organizations that share resources and knowledge. Governments and industry groups can play a role by offering subsidies or grants for organic certification and sustainable practices. Additionally, winemakers can focus on direct-to-consumer marketing strategies, emphasizing the authenticity and sustainability of their practices to justify a premium price and offset certification expenses.

Organic and natural winemaking must also contend with balancing sustainability goals with consumer preferences and market demands. While there is growing interest in eco-friendly products, not all consumers prioritize sustainability, and some may view the higher prices of organic and natural wines as a barrier.

Winemakers can address this by leveraging storytelling to communicate the value of their practices, including the environmental benefits and artisanal approach to winemaking. Engaging with wine education programs, hosting vineyard tours, and participating in sustainability-focused events can help build consumer loyalty. Offering a range of products, including more affordable options, can also broaden market appeal.

Through the adoption of innovative techniques, fostering collaboration, and engaging consumers, winemakers can navigate the challenges of organic and natural winemaking while staying true to their commitment to sustainability and quality.

Equipment and Tools for Small-Scale Production

Small-scale sustainable wine production involves the use of specialized equipment and tools tailored to the size of the operation while prioritizing environmentally friendly practices. Selecting appropriate equipment plays a vital role in producing high-quality wine, minimizing waste, and reducing energy consumption and environmental impact. Below is an overview of the essential tools and equipment used in small-scale sustainable winemaking.

Vineyard Management Tools are crucial for maintaining vine health and ensuring sustainable practices. Pruning shears and loppers are essential for managing canopy growth, and lightweight, ergonomic designs reduce labour fatigue. Tools like seed spreaders or small tillers are useful for managing cover crops, which contribute to soil fertility and erosion control. Small-scale compost turners are invaluable for recycling vineyard waste into nutrient-rich compost, enhancing soil health. Efficient drip irrigation systems with water-saving emitters ensure precise water delivery while minimizing wastage.

Figure 47: Wine grapes in the Italian wine region of Sicily being harvested by hand on the rocky soil. Fabio Ingrosso, CC BY 2.0, via Wikimedia Commons.

Harvesting Tools include reusable bins for collecting grapes and high-quality picking shears for selectively harvesting ripe clusters. For larger operations, portable mechanical harvesters provide an eco-friendly solution, reducing labour requirements while maintaining precision.

Figure 48: Pellenc 8390 Grape Harvester Machine. Pellencgroup, CC BY-SA 3.0, via Wikimedia Commons.

Crushing and Destemming Equipment ensures efficient preparation of grapes for fermentation. Hand-crank crushers are ideal for small-scale operations, providing gentle crushing without over-extraction. For added efficiency, crusher-destemmers combine the processes of crushing grapes and separating stems, with options for manual or motorized operation to suit different sustainability goals.

Fermentation Equipment is essential for converting grape juice into wine. Fermentation vessels, including stainless steel tanks, food-grade plastic barrels, or glass carboys, are chosen based on capacity and sustainability goals. Stainless steel is durable and easy to clean, while glass is perfect for small-batch fermentation. Airlocks and stoppers maintain an anaerobic environment to prevent contamination. Temperature control systems, such as cooling jackets or immersion probes, regulate fermentation temperatures, enhancing flavour profiles and quality.

Pressing Equipment like basket presses and bladder presses are used to extract juice or wine from crushed grapes. Basket presses offer a traditional, energy-efficient method suitable for small operations, while bladder presses use water pressure for precision and clarity.

Figure 49: A basket wine press being used on red wine grape varieties in Bordeaux. davitydave, CC BY 2.0, via Wikimedia Commons.

Clarification and Stabilization Tools ensure the wine's clarity and stability. Small-scale filtration systems, such as plate-and-frame or membrane filters, remove particles and microorganisms. Fining tools, including stirring rods or sieves, distribute agents like bentonite or activated carbon to achieve desired clarity and flavour.

Bottling and Sealing Equipment facilitates the final stage of winemaking. Manual bottling machines ensure precise and efficient filling, while manual or semi-automatic corkers seal bottles with natural, synthetic, or screw caps. Heat-shrink or manual capsuling devices add protective seals to bottles.

Cleaning and Sanitizing Tools are vital for maintaining hygiene in sustainable winemaking. Cleaning brushes and pads ensure thorough cleaning of tanks and equipment, while eco-friendly sanitizing solutions reduce environmental impact. Energy-efficient steam cleaners sanitize equipment without chemicals, aligning with sustainable practices.

Testing and Monitoring Instruments help maintain wine quality. Refractometers and hydrometers measure sugar levels (Brix) to monitor fermentation progress. pH meters and acid test kits ensure balanced acidity, while thermometers monitor fermentation and storage temperatures. Sulphur dioxide (SO_2) test kits help manage sulphite levels for preservation.

Aging and Storage Equipment includes oak barrels, stainless steel tanks, and glass carboys for aging wine. Oak barrels impart flavour complexity, while stainless steel offers durability and eco-friendliness. Simple humidity and temperature control systems ensure optimal cellar conditions for aging.

Waste Management Tools address the by-products of winemaking. Pomace composters recycle grape skins, seeds, and stems into organic fertilizers, while wastewater treatment systems filter effluents for reuse or safe discharge.

Renewable Energy Solutions enhance sustainability. Solar panels power equipment and lighting, while energy-efficient machinery like manual crushers and presses minimizes electricity consumption.

Aging, Bottling, and Packaging Sustainably

Sustainable wine aging, bottling, and packaging practices aim to minimize environmental impact while maintaining or enhancing wine quality. By integrating eco-friendly materials, energy-efficient methods, and innovative technologies, wineries can reduce their carbon footprint and appeal to environmentally conscious consumers. Below is a detailed explanation of sustainable practices in each stage.

Wine Aging Sustainably

Sustainable wine aging is an increasingly important focus in the wine industry, emphasizing the need to reduce energy consumption, minimize waste, and utilize eco-friendly materials. This synthesis examines the various sustainable practices in wine aging, particularly in relation to energy efficiency, aging vessels, and waste minimization.

Aging wine in temperature-controlled environments is essential for developing its flavour and aroma, but it can be energy-intensive. Sustainable wineries are adopting renewable energy sources, such as solar and geothermal power, to mitigate energy consumption during the aging process [401]. Additionally, innovative designs like earth-sheltered cellars and natural cooling systems can significantly reduce the energy required for temperature control [402]. The use of high-density polyethylene tanks has also been explored as a means to enhance energy efficiency while maintaining the quality of the wine [403].

Figure 50: A winery worker punches down the cap of red grape skins fermenting in an oak barrel. This process is known in French as pigeage and is part of the maceration process that extracts colour, flavour and aroma compounds from the grape skins into the wine. Ryan O'Connell, CC BY-SA 2.0, via Wikimedia Commons.

The choice of aging vessels plays a crucial role in sustainable wine production. Oak barrels, traditionally favoured for their ability to impart complex flavours, can be sourced sustainably or reconditioned to extend their lifespan, thereby reducing the demand for new timber [404]. Stainless steel tanks are another sustainable option; they are durable, recyclable, and require less energy for maintenance compared to wooden barrels [405]. Concrete vessels are gaining popularity due to their reusability and excellent thermal stability, which helps in maintaining optimal aging conditions without relying heavily on artificial temperature control [405]. Furthermore, the use of oak fragments instead of whole barrels allows for a reduction in aging time and costs while still imparting desirable flavour characteristics [406].

Effective waste management is a critical component of sustainable wine aging. Byproducts such as lees can be composted or repurposed for other vineyard uses, thus enhancing sustainability [407]. The recycling of wine lees not only reduces waste but also provides economic nutrients that can benefit vineyard health [407]. Moreover, the implementation of alternative aging technologies, such as micro-oxygenation, has been shown to improve the antioxidant activity and phenolic composition of wines, aligning with sustainability goals [408].

Sustainable Bottling Practices

Sustainable bottling practices in the wine industry are increasingly important as consumers and producers alike seek to minimize environmental impacts. This involves optimizing materials and processes to reduce waste and energy consumption while maintaining wine quality.

Eco-Friendly Bottles: One of the primary strategies for sustainable bottling is the use of lightweight glass bottles. These bottles significantly reduce energy consumption during both production and transportation due to their reduced weight, which leads to lower fuel usage in logistics [409]. Additionally, the use of recycled glass for bottle production can further decrease the demand for raw materials, thereby contributing to a more sustainable lifecycle for wine packaging [410]. The recycling of glass not only conserves resources but also minimizes waste, aligning with the principles of a circular economy [410].

Alternative Packaging Materials: Wineries are also exploring alternative packaging materials that can lower their carbon footprint. For instance, biodegradable or reusable materials such as bag-in-box formats or aluminium cans are gaining traction. These alternatives often have a lower environmental impact compared to traditional glass bottles [409]. Furthermore, innovations in bottle design, such as reducing wall thickness, can lead to significant reductions in material usage and energy consumption during production [410]. The adoption of these materials reflects a growing trend towards sustainability in the wine industry.

Efficient Bottling Processes: The efficiency of bottling processes is another critical aspect of sustainable practices. Automated bottling lines that utilize low energy consumption technologies can enhance operational sustainability [411]. The integration of renewable energy sources into these processes further amplifies their environmental benefits [411]. Moreover, refillable bottle systems are being introduced, promoting reuse and reducing the need for single-use packaging [411]. Such systems not only decrease waste but also foster a culture of sustainability among consumers.

Caps and Closures: The choice of caps and closures also plays a significant role in sustainable bottling. Natural corks, harvested from renewable cork oak trees, are biodegradable and have a lower environmental impact compared to synthetic alternatives [411]. On the other hand, screw caps, while made from aluminium, are lightweight and recyclable, making them a practical choice when paired with effective recycling programs [411]. The sustainability of closures is crucial, as they contribute to the overall environmental footprint of wine packaging.

Sustainable Packaging Solutions

Sustainable packaging solutions are increasingly critical in the wine industry, as they directly influence environmental impact through waste reduction, improved recyclability, and lower transportation emissions. The adoption of sustainable packaging practices is essential for wineries aiming to align with global sustainability goals and consumer preferences for eco-friendly products.

Recyclable and Biodegradable Materials: The use of recyclable and biodegradable materials, such as cardboard made from recycled paper, biodegradable inks, and compostable shipping cartons, is fundamental in promoting eco-friendliness in wine packaging. Research indicates that packaging contributes significantly to the overall carbon footprint of wine production, with glass bottles accounting for approximately 30% to 40% of the total carbon footprint associated with wine distribution when considering the entire production process and transport emissions [412]. By utilizing sustainable materials, wineries can mitigate these impacts and enhance their environmental credentials [413].

Lightweight Shipping Materials: The concept of lightweighting shipping materials is vital for reducing fuel consumption and greenhouse gas emissions during transportation. Studies show that optimizing the thickness of glass and the overall size and weight of packaging can lead to significant energy savings and reduced emissions [414]. This approach not only lowers costs associated with transportation but also aligns with broader sustainability initiatives within the industry.

Alternative Packaging Options: Innovative packaging solutions such as Bag-in-Box (BiB) and aluminium cans are gaining traction in the wine industry. Bag-in-Box packaging is particularly effective for bulk wine distribution, as it is lightweight and recyclable, thus reducing shipping emissions [413]. Similarly, aluminium cans are fully recyclable and have become an increasingly popular alternative to traditional glass bottles, appealing to environmentally conscious consumers. The shift towards these alternatives reflects a broader trend in the industry towards sustainable practices that meet consumer demand for eco-friendly options [415].

Reusable and Returnable Systems: Implementing reusable and returnable packaging systems, such as reusable glass bottles or kegs for bulk wine sales, can significantly reduce single-use waste. This approach not only minimizes environmental impact but also fosters a circular economy within the wine sector. Research highlights that such systems can enhance sustainability performance by promoting resource efficiency and reducing the overall carbon footprint of wine distribution [416].

Certifications and Consumer Transparency in Sustainable Winemaking

Certifications and transparency are essential in showcasing a winery's commitment to sustainability. Certifications establish verifiable standards, holding producers accountable for their environmental and social practices. Transparency ensures that customers are well-informed about these efforts, building trust, enhancing brand reputation, and fostering consumer loyalty. Together, they create a foundation for responsible winemaking that appeals to environmentally conscious buyers.

Certifications in sustainable winemaking act as benchmarks for ecological and social responsibility. They verify that a winery's operations adhere to strict criteria, covering aspects like vineyard management, bottling, and packaging. LEED (Leadership in Energy and Environmental Design) certification is awarded to facilities that demonstrate energy efficiency, water conservation, waste reduction, and sustainable building practices. For wineries, achieving LEED certification often involves using renewable energy systems, reducing water usage, and designing energy-efficient infrastructure, signalling to consumers a commitment to environmental stewardship. B Corp Certification evaluates a winery's social and environmental performance, transparency, and accountability. A B Corp-certified winery demonstrates responsible practices in employee welfare, community involvement, and ecological impact, balancing profit with purpose. Carbon Neutral Certifications acknowledge wineries that measure, reduce, and offset their carbon emissions. These wineries adopt practices like renewable energy use, optimizing supply chains, and purchasing carbon offsets to mitigate their greenhouse gas footprint, ensuring a carbon-neutral operation.

Transparency complements certifications by connecting these verified practices to consumer awareness. This involves clear communication, detailed labelling, and engaging storytelling about the winery's sustainability journey. Labels provide a direct avenue to inform consumers about certifications and eco-friendly initiatives. For example, a wine bottle may display logos for LEED or Carbon Neutral certifications with accompanying explanations. Labels can also highlight sustainable features like recycled packaging or lightweight bottles. Marketing and storytelling further enhance transparency by creating an emotional connection with consumers. Wineries can use digital platforms, newsletters, or tasting room experiences to share their efforts in water conservation, renewable energy, or organic farming. Traceability extends transparency by providing detailed insights into wine production, from vineyard to bottle. Consumers increasingly seek information about grape origins, farming methods, and production processes. QR codes and traceability platforms make this information easily accessible, deepening consumer trust.

The benefits of certifications and transparency are many. They build consumer trust and loyalty by validating sustainability claims, fostering long-term customer relationships. Certifications also differentiate wineries in a competitive market where demand for sustainable products is rising. Transparency educates consumers on the importance of sustainability in winemaking, inspiring environmentally conscious purchasing decisions. By adopting certifications, wineries contribute to raising industry-wide standards, promoting a culture of responsibility.

Despite these advantages, challenges persist. The certification process can be complex, requiring extensive documentation, audits, and significant costs. Small wineries can address these challenges by seeking grants or partnering with organizations that support sustainable initiatives. Consumer misunderstanding of certification labels poses another obstacle, which wineries can overcome through clear, concise explanations on labels and marketing materials. Balancing transparency with branding requires careful messaging. Overly technical details can overwhelm consumers, so wineries should focus on simplified, impactful communication that highlights key achievements.

Chapter 9
Marketing and Selling Sustainable Wine

Building Your Brand Around Sustainability

Building a wine brand centred around sustainability requires a multifaceted approach that integrates environmental responsibility into every aspect of the business. This involves establishing core sustainability values, implementing eco-friendly practices, obtaining certifications, and effectively communicating these efforts to consumers.

The foundation of a sustainable wine brand is a clear commitment to environmental and social responsibility. This includes reducing carbon emissions, conserving water, promoting biodiversity, and ensuring fair labour practices. Research indicates that consumers with positive attitudes towards the environment are willing to pay more for sustainable wines, suggesting that a strong commitment to sustainability can enhance brand loyalty and consumer trust [417, 418]. Furthermore, embedding these values into vineyard management and production processes is essential for authenticity and consumer perception [90, 176].

Adopting sustainable practices in both the vineyard and winery is crucial for reinforcing brand authenticity. Organic or biodynamic farming, the use of cover crops, and drip irrigation are effective vineyard practices that contribute to sustainability [419, 420]. In the winery, energy-efficient machinery and renewable energy sources can significantly reduce environmental impact [91, 421]. Additionally, sustainable packaging options, such as lightweight bottles and recycled materials, further demonstrate a commitment to reducing carbon footprints [422, 423].

Certifications serve as a validation of a wine brand's sustainability claims, enhancing credibility among consumers. Certifications like Organic, Biodynamic, and B Corp provide verifiable benchmarks of environmental responsibility [91, 424]. The presence of these certifications on labels can significantly influence consumer purchasing decisions, as they assure consumers of the brand's commitment to sustainable practices [419, 423].

Transparency in communication is vital for connecting with consumers. Utilizing labels, websites, and marketing campaigns to highlight sustainability efforts can build trust and engagement [425, 426]. For instance, QR codes on bottles can direct consumers to detailed information about the vineyard's sustainable practices, enhancing their understanding and appreciation of the brand [82, 418].

Storytelling is an effective way to personalize a brand and make sustainability relatable. Sharing narratives about the vineyard's journey towards sustainability, including challenges and successes, can resonate with consumers [426, 427]. Highlighting the people behind the brand, such as winemakers and vineyard workers, can also enhance emotional connections with consumers [90, 176].

Collaborating with environmental organizations and local communities can amplify sustainability efforts. Participating in events like vineyard clean-ups or sustainability workshops reinforces a brand's dedication to environmental stewardship [420, 423]. Such partnerships not only enhance credibility but also expand the brand's reach within the community [176, 422].

Educating consumers about sustainable winemaking practices can deepen their appreciation for the brand. Providing accessible content about the benefits of organic farming and water conservation can engage consumers and foster loyalty [421, 422]. Interactive experiences, such as winery tours or virtual tastings, can further enhance consumer understanding and engagement [91, 424].

The visual identity of a brand should reflect its sustainable ethos. Utilizing eco-friendly materials for labels and packaging, along with imagery that conveys environmental commitment, can strengthen brand recognition and consumer loyalty [90, 425]. This visual representation can serve as a powerful marketing tool to attract environmentally conscious consumers [91, 423].

Encouraging satisfied customers to share their experiences can build trust and attract like-minded consumers. User-generated content, reviews, and testimonials showcasing the wine's quality and eco-friendliness can enhance brand credibility [424, 427]. Social media campaigns can further amplify the brand's sustainability message and engage a broader audience [418, 421].

Sustainability is an evolving field, and brands must regularly assess their practices and seek innovations to reduce environmental impact [420, 421]. Staying informed about industry trends and consumer preferences is crucial for maintaining a competitive edge in sustainable winemaking [176, 428].

Building a wine brand around sustainability not only fulfills a moral imperative but also provides a strategic advantage in a market increasingly focused on ethical consumption. A strong commitment to sustainability differentiates a brand, attracts loyal customers, and positions the wine as a product that aligns with both quality and values [90, 417, 418]. By embedding sustainability into every facet of operations and communicating these efforts effectively, a wine brand can resonate with conscious consumers and contribute positively to the environment and society.

Direct-to-Consumer Models

Selling Wine

Laws and regulations for selling wine by small-scale producers vary significantly across the globe, reflecting the diverse legal, cultural, and economic landscapes of the wine industry. These rules encompass production standards, licensing, labelling, taxation, distribution, and marketing. Understanding and complying with these regulations is essential for small-scale winemakers to ensure legal operations and foster sustainable growth.

Small-scale producers must typically secure specific licenses or permits to produce and sell wine. In the United States, winemakers are required to obtain a Basic Permit from the Alcohol and Tobacco Tax and Trade Bureau (TTB), along with any necessary state and local permits. In the European Union, producers must register with local agricultural authorities and adhere to EU-wide wine production regulations, such as those governing geographic indications (GI). In countries like Australia, wineries must obtain a producer's license and comply with the Wine Equalisation Tax (WET) system, while in New Zealand, a winemaker's license is mandatory under the Sale and Supply of Alcohol Act. These licensing requirements establish a foundational framework for lawful wine production and sales.

Labelling regulations are designed to protect consumers and preserve regional identities. Wine labels must often display critical information such as alcohol content, volume, producer details, and health warnings. Geographic indications (GIs), like Champagne in France or Barossa Valley in Australia, require producers to meet specific regional standards to use these protected terms. Organic or sustainable certifications also necessitate compliance with stringent requirements and appropriate labelling to differentiate these products in the marketplace.

Taxation and excise duties significantly influence profitability for small-scale producers. In the United States, federal excise taxes apply, with reduced rates available for small producers, while state taxes vary widely. The European Union applies excise duties at the member-state level, offering exemptions in some cases. In Australia, the value-based Wine Equalisation Tax (WET) includes rebates for small wineries, making compliance manageable for smaller operations. These tax systems reflect the importance of balancing fiscal responsibility with industry support.

Direct-to-consumer (DTC) sales regulations vary by region and offer opportunities for small-scale producers to connect with customers. In the United States, DTC sales are permitted in many states but require navigating complex shipping and licensing laws. In the European Union, producers can sell directly at their wineries or through local markets, though cross-border sales may require additional documentation. In Australia and New Zealand, DTC sales, including online and cellar-door options, are widely permitted but may necessitate specific licenses.

Export regulations further complicate the landscape for small producers aiming to access international markets. In the United States, exporting wine requires an Export Certificate from the TTB, alongside adherence to destination-country import laws. EU producers must meet export standards and provide documentation such as VI-1 certificates for non-EU markets. In Australia, Wine Australia oversees export licensing and compliance with international trade agreements. Understanding these requirements is crucial for expanding into global markets.

Health and safety standards are essential for wine production. Facilities must meet food-grade sanitation and hygiene standards, while laws regulate the use of additives such as sulphur dioxide. Regular testing for parameters like alcohol content, acidity, and microbial stability ensures compliance and product quality. These standards safeguard consumer health while reinforcing trust in the product.

Marketing and advertising restrictions aim to promote responsible consumption. In the United States, advertising must avoid targeting minors or making misleading health claims. EU marketing laws must comply with both EU-wide and national regulations, while countries like Australia and New Zealand regulate advertising to ensure responsible messaging that does not appeal to underage consumers.

Small-scale producers face several challenges, including the complexity of compliance with overlapping regulations, high costs associated with licensing, taxes, and testing, and restricted market access due to DTC limitations or limited distribution networks. Additionally, exporting wine can involve significant costs and paperwork. Producers can address these challenges by seeking legal or consultancy advice, leveraging grants or subsidies for small businesses, and forming partnerships with local wine clubs or online platforms. Utilizing trade associations or export assistance programs can simplify international trade.

Global trends in wine regulations highlight a shift toward sustainability and innovation. Many regions now offer tax incentives or grants to encourage sustainable practices, while the increasing acceptance of online wine sales has spurred new e-commerce licensing frameworks. Simplified compliance schemes, such as rebates for small producers in Australia, further support small-scale wineries in achieving their business objectives.

Navigating the diverse regulatory landscape is a critical aspect of small-scale winemaking. By staying informed, leveraging support networks, and adopting best practices, producers can successfully balance compliance with operational and financial goals, ensuring the longevity and success of their ventures in an evolving industry.

Farm-to-Table Partnerships

Farm-to-table partnerships in winemaking represent collaborative efforts between small-scale wineries, local farmers, restaurants, and consumers to establish a direct connection between agricultural producers and the end consumer. These partnerships prioritize sustainability, community engagement, and the promotion of high-quality, locally sourced products. For wineries, such collaborations enhance the story of their wine, tying it to the local terroir and agricultural heritage while creating economic opportunities and fostering environmental stewardship.

These partnerships focus on the use of locally grown grapes and agricultural products in winemaking and related activities. Extending beyond the winery itself, they involve local food producers, chefs, and restaurants, promoting a holistic approach to sustainability and quality. This concept champions the belief that food and beverages should be produced, sourced, and consumed locally whenever possible, reducing the environmental impact associated with transportation and industrial farming.

Farm-to-table partnerships provide numerous benefits for wineries. By sourcing grapes and other raw materials locally, wineries ensure fresh, high-quality inputs that reflect the unique characteristics of their region. Tying the wine to its local environment creates a compelling narrative that resonates with consumers seeking authenticity and transparency. Additionally, local sourcing reduces transportation emissions and supports eco-friendly farming practices, contributing to the winery's sustainability goals. These partnerships stimulate local economies by fostering goodwill and strong community ties. Collaborations with local farmers can also inspire innovative wine pairings or the use of unique, locally grown ingredients in specialty wines or beverages, diversifying the winery's offerings.

Establishing successful farm-to-table partnerships involves collaboration, clear communication, and shared values. Wineries can partner with local grape growers, farmers, and suppliers who align with sustainable practices. These agreements may include sourcing organic or biodynamic grapes or using locally grown herbs, honey, or fruits for specialty wines. Collaborations with chefs who prioritize local ingredients can result in exclusive wine pairings or menus, while vineyard tours, harvest festivals, and "meet the maker" events can showcase the connection between the winery and local agriculture. Promoting local products in tasting rooms or winery stores and educating consumers about the importance of local sourcing through storytelling, labelling, and digital platforms can build trust and loyalty.

Challenges such as logistical coordination, limited local supply, and consumer education often arise in these partnerships. Addressing these issues requires careful supply chain management and collaboration with multiple local suppliers to ensure consistent quality and availability. For wineries with growing demand, maintaining a balance between local sourcing and scalability is critical, and gradual expansion of partnerships along with diversified offerings can help. Transparent labelling, storytelling, and marketing can bridge knowledge gaps, helping consumers understand the environmental and quality benefits of farm-to-table partnerships.

Sustainability is central to farm-to-table partnerships, aligning closely with sustainable winemaking practices. By fostering local relationships, wineries support biodiversity, reduce waste, and promote organic or regenerative farming methods. These efforts contribute to broader sustainability goals such as reducing carbon footprints and supporting resilient local food systems.

Farm-to-table partnerships enhance the value proposition of small-scale wineries by emphasizing local sourcing, community involvement, and sustainability. These partnerships serve as both an economic and environmental strategy, while also deepening connections with consumers and celebrating the rich agricultural traditions of a region. By investing in local relationships and promoting their stories, wineries can position themselves as leaders in the farm-to-table movement, delivering exceptional wines with a meaningful purpose.

Hosting Vineyard Tours and Tastings for Small-Scale Wineries

Small-scale wineries can benefit greatly from hosting vineyard tours and tastings, which serve as an effective way to engage consumers, build brand loyalty, and create memorable experiences. These events allow wineries to showcase their unique production methods, sustainable practices, and the quality of their wines while fostering a personal connection between visitors and the vineyard.

Hosting vineyard tours and tastings is not just about offering wine samples but about providing a comprehensive experience that educates and entertains visitors. For small-scale wineries, this is an opportunity to differentiate themselves from larger operations through personal interactions and storytelling. These events allow visitors to learn about the winemaking process, from grape cultivation to bottling, and to appreciate the nuances of the wine's flavour and aroma profiles.

Tastings and tours also serve as a powerful marketing tool. Visitors who experience the vineyard firsthand are more likely to become loyal customers, purchase wines directly, and recommend the winery to others. Additionally, these events can generate supplementary income through direct sales, exclusive tastings, or wine club memberships.

A well-structured vineyard tour offers a balance of education, interaction, and enjoyment. The tour typically begins with a walk through the vineyard, where visitors can observe grape varieties, vine training methods, and sustainable practices. Guides explain the significance of terroir—the unique combination of soil, climate, and cultivation techniques that influence the wine's character. Sharing anecdotes about the vineyard's history or the winemaker's journey can add a personal touch.

The tour then proceeds to the production area, where visitors can see winemaking equipment and learn about fermentation, pressing, aging, and bottling processes. Highlighting sustainable practices, such as energy-efficient equipment or organic farming, can further enhance the visitor's appreciation of the winery's commitment to environmental stewardship.

The tasting portion is often the highlight of the visit, allowing guests to sample a curated selection of wines. Tastings may include vertical tastings (different vintages of the same wine) or horizontal tastings (wines of the same vintage but different varietals). Wineries can offer guided tastings led by knowledgeable staff or the winemaker, explaining the flavour profiles, aroma notes, and food pairings for each wine.

To create an engaging experience, wineries can provide tasting notes, wine pairing recommendations, or sensory activities, such as aroma identification. For small-scale operations, offering unique wines or limited-edition releases during tastings can create exclusivity and encourage purchases.

The ambiance of the tasting room and vineyard plays a critical role in the overall experience. A comfortable and aesthetically pleasing setting, with well-arranged seating, scenic views, and thoughtful décor, can elevate the experience. Personal interaction is key in small-scale wineries; engaging with visitors on a personal level fosters connection and leaves a lasting impression.

Hosting vineyard tours and tastings comes with challenges, such as staffing, infrastructure, and maintaining quality during peak seasons. Small-scale wineries can overcome these challenges by hiring passionate, well-trained staff who can deliver exceptional customer service. Offering pre-booked tours can help manage visitor flow and ensure a more personalized experience.

Infrastructure upgrades, such as shaded seating areas or accessible pathways, can improve guest comfort without requiring large-scale investments. Additionally, wineries can partner with local businesses for catering or transportation services to enhance the overall experience.

To attract visitors, small-scale wineries can use social media, email newsletters, and partnerships with tourism boards or local businesses. Promoting events like seasonal harvest festivals or exclusive wine release parties can drive attendance. Encouraging satisfied visitors to share their experiences through reviews and social media can further amplify the winery's reach.

Vineyard tours and tastings are an invaluable tool for small-scale wineries to showcase their craftsmanship, connect with consumers, and build brand loyalty. By creating engaging, informative, and enjoyable experiences, wineries can turn visitors into ambassadors who appreciate and support their unique wines and sustainable practices. With careful planning and attention to detail, hosting tours and tastings can become a cornerstone of a small-scale winery's success.

Leveraging Online and Social Media Platforms

In the context of small-scale wineries, leveraging online and social media platforms has become essential for expanding market reach, enhancing sales, and fostering customer engagement. These digital tools allow wineries to effectively showcase their products, narrate their unique stories, and cultivate a loyal customer base in a cost-efficient manner. The strategic implementation of these platforms enables small wineries to compete with larger brands and carve out a distinctive identity in the digital marketplace.

Establishing a robust online presence is crucial for small wineries, starting with a well-designed website that acts as a digital storefront. The website should include high-quality visuals of the vineyard and wine offerings, alongside essential functionalities such as an intuitive online store and secure payment options. Detailed information about the winery's history, sustainability practices, and product descriptions is also vital for building consumer trust and interest [429, 430]. Incorporating a blog or news section can further enhance engagement by allowing wineries to share updates, events, and educational content, such as wine-pairing tips or insights into the winemaking process. Search engine optimization (SEO) is another critical aspect, as it helps attract organic traffic, ensuring that potential customers can easily discover the winery when searching for relevant terms [429, 431].

Social media platforms like Instagram, Facebook, TikTok, and Pinterest present unique opportunities for wineries to connect with their audience. Each platform serves different demographics and user behaviours, necessitating tailored content strategies. For instance, Instagram is particularly effective for visual storytelling, showcasing the vineyard's beauty and the craftsmanship behind the wines through high-quality images and engaging videos [432]. Facebook serves as a versatile platform for longer posts and community building, while TikTok offers a space for creative, short-form content that can attract younger consumers [433]. Pinterest can be utilized for sharing aesthetically pleasing content, such as wine-pairing recipes, which can drive traffic back to the winery's website [432]. By sharing authentic narratives about their journey from vine to bottle, introducing team members, and highlighting sustainable practices, wineries can foster deeper connections with their audience [432, 434].

The integration of e-commerce capabilities allows wineries to sell directly to consumers, bypassing traditional distribution channels. User-friendly online sales platforms should feature detailed product descriptions, various shipping options, and recommendations for wine pairings or bundles [432]. Additionally, social media platforms increasingly incorporate shopping features, such as Instagram Shopping and Facebook Marketplace, enabling wineries to showcase their products directly to consumers [432]. Linking these features to the main website simplifies the purchasing process, potentially increasing conversion rates and driving sales.

Email marketing remains a powerful tool for nurturing customer relationships and driving repeat sales. Wineries can utilize email newsletters to keep subscribers informed about new product releases, upcoming events, and exclusive discounts. Personalizing emails based on past purchases or preferences enhances the customer experience, fostering loyalty and encouraging repeat business [432, 434].

Collaborations with influencers, bloggers, or local chefs can significantly amplify a winery's online presence. Influencers can provide authentic reviews and creative content that introduces the winery to a broader audience [435]. Partnering with complementary businesses, such as gourmet food suppliers or event planners, can also expand the winery's reach and enhance its visibility in the market [435].

The COVID-19 pandemic accelerated the adoption of virtual wine tastings, allowing wineries to connect with customers globally. Small-scale wineries can organize online tasting sessions where participants receive wine samples in advance, enabling winemakers to guide them through tasting notes and food-pairing suggestions [432, 434]. These virtual events not only promote wine sales but also help build rapport with customers in a unique and engaging manner.

Digital platforms provide valuable analytics tools to measure the effectiveness of online campaigns. Metrics such as website traffic, social media engagement, and conversion rates offer insights into audience preferences and behaviours [431, 432]. Regular analysis allows wineries to refine their strategies, focusing on the most successful platforms and content types to maximize their marketing efforts.

Despite the potential benefits of digital platforms, small-scale wineries may encounter challenges such as limited budgets and competition from larger brands. These challenges can be addressed through cost-effective content creation, utilizing smartphone technology for high-quality visuals, and leveraging free or low-cost graphic design tools [431, 434]. Education and training through online resources or workshops can enhance marketing skills, while focusing on authenticity and personal connections can help small wineries differentiate themselves from mass-market brands [431, 434].

Certifications and Labels (Organic, Biodynamic, etc.)

Wine certifications and labels serve as official endorsements that validate a wine's adherence to specific standards related to quality, origin, production practices, and sustainability. They offer essential information to consumers and industry stakeholders, acting as a symbol of credibility and trust. These certifications and labels span categories such as geographical indications, organic production, sustainability, ethical practices, and food safety, ensuring a comprehensive framework for assessing wine quality and production integrity.

The primary purpose of wine certifications and labels is to assure consumers that the wine meets specific quality, origin, or production standards, instilling confidence in their purchase decisions. Labels provide transparency by offering clear, regulated information about a wine's origin, varietals, production methods, and environmental credentials. They enable market differentiation, helping wines stand out in a competitive industry by appealing to consumers who value sustainability, authenticity, or unique production methods. Certifications also ensure compliance with national and international standards, particularly in export markets, and enhance brand credibility by demonstrating the winery's commitment to quality, sustainability, and ethical practices.

For small-scale wineries, certifications and labels can be transformative tools. While the certification process may involve costs and administrative work, the benefits significantly outweigh these challenges. Certifications grant small wineries access to niche markets, allowing them to compete effectively with larger producers. For instance, certifications like Organic or Biodynamic appeal to health-conscious and environmentally aware consumers, while Geographical Indications (GIs) protect a wine's regional identity in global markets.

These certifications also foster consumer appeal and loyalty. Modern consumers increasingly seek transparency in their purchasing choices, and certifications such as Fair Trade, Carbon Neutral, or Sustainable Wine resonate strongly with those prioritizing ethical and sustainable consumption. Highlighting certifications on labels creates a deeper emotional connection with the brand. Furthermore, certified wines often command higher prices, as they are perceived as higher quality or more ethically produced. Labels such as DOCG (Italy) or Certified Sustainable justify premium pricing for artisanal wines.

Certifications also act as third-party endorsements, bolstering credibility and trust for small wineries without the extensive marketing budgets of large-scale competitors. They demonstrate adherence to rigorous quality and production standards, building trust among distributors, retailers, and consumers. Additionally, certifications help wineries navigate regulatory landscapes, especially in export markets. For instance, EU wines must comply with strict labelling laws, and certifications like PDO (Protected Designation of Origin) facilitate compliance with such regulations.

Alignment with sustainability trends is another significant advantage. Certifications like LEED, Sustainable Winegrowing New Zealand, or Carbon Neutral align small wineries with global sustainability movements, enhancing brand reputation and contributing to environmental responsibility. Labels and certifications also add value to a winery's storytelling and marketing efforts by tying the wine to a specific region, tradition, or value system. For example, certifications like Slow Wine or Demeter Biodynamic highlight craftsmanship and heritage in branding efforts.

Finally, certifications offer an opportunity for consumer education and differentiation. They allow small-scale wineries to educate customers about what sets their wines apart, whether through labels like Vegan Wine or Natural Wine that attract specific consumer groups or by showcasing their unique production philosophies. These efforts foster a deeper appreciation for the wine and create loyal, informed customers.

The following provides a list of various wine certifications and labels from around the world, encompassing quality standards, geographical indications, sustainable practices, and organic certifications. These certifications ensure transparency, quality, and adherence to environmental or traditional standards.

Geographical Indications (GI) and Quality Standards

These certifications indicate that a wine originates from a specific region and adheres to established production standards.

Europe

- **Appellation d'Origine Contrôlée (AOC)** – France
- **Denominazione di Origine Controllata (DOC)** – Italy
- **Denominazione di Origine Controllata e Garantita (DOCG)** – Italy
- **Indicazione Geografica Tipica (IGT)** – Italy
- **Denominación de Origen (DO)** – Spain
- **Denominación de Origen Calificada (DOCa)** – Spain
- **Vino de Pago (VP)** – Spain
- **Qualitätswein (QbA)** – Germany
- **Prädikatswein** – Germany (includes Kabinett, Spätlese, Auslese, Beerenauslese, Trockenbeerenauslese, and Eiswein)
- **Vin de Pays (IGP)** – France
- **PDO (Protected Designation of Origin)** – European Union
- **PGI (Protected Geographical Indication)** – European Union

New World

- **American Viticultural Areas (AVA)** – United States
- **Geographical Indications (GI)** – Australia
- **Wine of Origin (WO)** – South Africa
- **Denomination of Origin (DO)** – Chile
- **Geographical Indications (GI)** – Canada
- **Indicaciones Geográficas Protegidas (IGP)** – Argentina

Sustainability Certifications

These labels ensure adherence to environmentally friendly and socially responsible winemaking practices.

- **Certified California Sustainable Winegrowing (CCSW)** – United States

- **Sustainability in Practice (SIP)** – United States

- **LIVE Certified (Low Input Viticulture and Enology)** – United States (Oregon, Washington)

- **Salmon-Safe Certification** – United States (Pacific Northwest)

- **Sustainable Wine of Chile** – Chile

- **Certified Sustainable Wine of South Africa** – South Africa

- **Sustainable Winegrowing New Zealand (SWNZ)** – New Zealand

- **EntWine Certification** – Australia

- **Fair'n Green** – Germany

- **Terra Vitis** – France

Organic Certifications

Organic wine certifications ensure that grapes are grown without synthetic pesticides, herbicides, or fertilizers, and that winemaking practices follow organic principles.

- **USDA Organic** – United States

- **European Union Organic Wine (EuroLeaf)** – European Union

- **Australia Certified Organic (ACO)** – Australia

- **Canada Organic Regime (COR)** – Canada

- **Bio Suisse** – Switzerland

- **Demeter Biodynamic Certification** – Global

- **AB (Agriculture Biologique)** – France

- **Naturland** – Germany

- **BioGro** – New Zealand

Biodynamic Certifications

Biodynamic certifications ensure adherence to practices that integrate ecological and spiritual principles in vineyard management.

- **Demeter International Certification** – Global
- **Biodyvin** – Europe
- **Biodynamic Association Certification** – United States

Natural Wine Certifications

While natural wines are less regulated, some associations have established certification systems.

- **Vin Méthode Nature** – France
- **Raw Wine Certification** – Global (for wines showcased at RAW Wine fairs)
- **SICAW (Spanish Independent Certified Artisan Wine)** – Spain

Fair Trade and Social Responsibility Certifications

These certifications focus on fair labour practices and support for local communities.

- **Fair Trade Certified** – Global
- **Fair for Life** – Global
- **Fairtrade International** – Global
- **WIETA Ethical Trade Certification** – South Africa

Climate and Carbon Certifications

These labels recognize efforts to minimize carbon footprints and combat climate change.

- **Carbon Neutral Certification** – Global
- **Climate Neutral Certification** – Global
- **Carbon Trust Standard** – United Kingdom

- **Certified B Corporation (B Corp)** – Global

Kosher and Religious Certifications

These certifications indicate compliance with religious dietary laws.

- **Kosher Certification** – Global (includes Orthodox Union, Star-K, and others)
- **Halal Certification** – For non-alcoholic wine products in Muslim-majority countries

Health and Safety Certifications

These certifications ensure that wines meet health and safety standards for consumption.

- **ISO 22000** – International Food Safety Management Systems
- **HACCP (Hazard Analysis and Critical Control Points)** – Global
- **IFS (International Featured Standard)** – Europe

Other Specialty Certifications

These certifications cater to unique winemaking practices and consumer interests.

- **Vegan Wine Certification** – Global (Vegan Society or Certified Vegan)
- **Gluten-Free Certification** – For wines confirmed to be free of gluten
- **Organic Biodynamic Vegan Wine (OBV)** – Combined certification by some wineries
- **Non-GMO Project Verified** – For wines made without genetically modified organisms
- **Slow Wine Movement Certification** – Italy, based on Slow Food principles
- **Zero-Waste Certification** – Recognizing wineries with waste reduction initiatives

Regional Certifications

Some countries or regions have additional certifications unique to their winemaking traditions.

- **Vinho Verde Certification** – Portugal

- **Tokaji Aszú Classification** – Hungary

- **Chianti Classico Gallo Nero** – Italy

- **Barossa Trust Mark** – Australia

- **Napa Green Certified Winery** – United States (Napa Valley)

This list highlights the diversity of certifications and labels, reflecting the complexity and global scope of wine production. Each certification ensures adherence to specific standards, enabling wineries to meet consumer expectations for quality, sustainability, and authenticity.

Chapter 10

Challenges and Opportunities in Sustainable Viticulture

Common Challenges for Small-Scale Growers

Small-scale wine growers encounter a variety of challenges that significantly impact their operations and viability. These challenges stem from economic constraints, environmental factors, pest and disease management, labour shortages, market competition, regulatory compliance, quality maintenance, sustainability pressures, and consumer education. Each of these areas presents unique hurdles that require innovative strategies and collaborative efforts to overcome.

Financial limitations are a primary concern for small-scale wine growers. The initial investment required for establishing a vineyard, including land acquisition, equipment, and labour, can be substantial. Many small producers struggle to secure affordable financing, which restricts their ability to invest in necessary upgrades or expansions [436]. Additionally, the volatility of grape prices and the lack of economies of scale further complicate profitability [437]. To navigate these economic challenges, small-scale growers often pursue grants, subsidies, or partnerships with local organizations, and they may adopt community-supported agriculture (CSA) models to share resources and reduce costs [438].

The wine industry is particularly vulnerable to climate change and extreme weather events, which can drastically affect grape quality and yield. Factors such as frost, drought, and excessive rainfall pose significant risks to vineyard health [439]. Small-scale growers can mitigate these risks by employing sustainable agricultural practices, such as planting drought-resistant grape varieties and utilizing precision agriculture technologies to monitor environmental conditions [440]. These adaptations are essential for maintaining productivity in the face of changing climate patterns [439].

Pests and diseases remain a constant threat to vineyard health, with common issues including infestations by grapevine moths and fungal diseases like powdery mildew [441]. Small-scale growers often lack the resources to implement comprehensive pest management strategies, making their crops more susceptible to damage [442]. Integrated Pest Management (IPM) offers a sustainable approach that combines biological controls and cultural practices to enhance vineyard resilience [442]. By adopting IPM, small growers can improve their pest management while minimizing environmental impact.

Labour availability is another critical challenge for small-scale wine producers, particularly during peak seasons such as planting and harvest [436]. The high cost of mechanization often makes it impractical for smaller operations, leading to reliance on seasonal labour, which can be difficult to secure [443]. To address labour shortages, some growers engage in community volunteer programs or labour-sharing agreements with neighbouring farms, while investing in training programs to develop a skilled local workforce [438].

Small-scale wine growers face intense competition from larger producers with established distribution networks and marketing budgets [443]. Competing effectively requires small producers to carve out niche markets by emphasizing the unique qualities of their wines, such as organic or biodynamic certifications [444]. Direct-to-consumer sales strategies, including online platforms and local farmers' markets, can help small-scale growers reach their target audiences more effectively [443].

Navigating the complex regulatory landscape of wine production can be daunting for small-scale growers. Compliance with labelling, taxation, and health standards is often time-consuming and costly, particularly when regulations vary across regions [445]. Joining industry associations or cooperatives can provide valuable support in understanding and managing these requirements [445]. Additionally, digital tools designed for vineyard management can streamline compliance processes, allowing growers to focus more on production [445].

Achieving consistent quality across vintages is a significant challenge for small-scale wine producers, influenced by variations in climate and winemaking techniques [436]. Implementing quality control measures, such as laboratory testing and advanced fermentation monitoring, can help ensure product consistency [436]. Flexibility in winemaking practices, such as blending different grape varieties, allows producers to adapt to fluctuations in grape quality [436].

Increasing consumer and regulatory demands for sustainable practices place additional pressure on small-scale growers [440]. Implementing eco-friendly practices often requires upfront investments that can strain limited budgets [446]. However, obtaining certifications like organic or biodynamic can enhance marketability and justify these costs [440]. Collaborating with local sustainability initiatives and applying for grants focused on environmental stewardship can also provide financial support [440].

Finally, many consumers are unaware of the unique challenges faced by small-scale wine producers, which can hinder brand loyalty and premium pricing [444]. Educating consumers about the artisanal nature of their products and the stories behind them is crucial for building a loyal customer base [444]. Utilizing social media and engaging with local restaurants and wine shops can help raise awareness and appreciation for small-scale wines [444].

Overcoming Barriers with Innovation

Winemaking is a complex endeavour that faces numerous barriers, including production inefficiencies, sustainability challenges, market access issues, and consumer engagement difficulties. Innovation is essential for winemakers, particularly small-scale producers, to navigate these obstacles effectively. By leveraging modern technologies, sustainable practices, and creative marketing strategies, winemakers can enhance their operational efficiency, product quality, and market presence.

One of the primary challenges in winemaking is optimizing production while managing costs and resources. Precision agriculture technologies, such as drones and IoT-enabled sensors, allow winemakers to monitor vineyard conditions in real time, facilitating targeted interventions that conserve resources and enhance productivity

[447, 448]. For instance, UAVs (Unmanned Aerial Vehicles) can assist in assessing vineyard health and identifying areas needing attention, thus improving overall vineyard management efficiency [449]. Additionally, automated equipment like mechanical harvesters can alleviate labour shortages and improve operational efficiency, particularly for small-scale producers who may benefit from compact, cost-effective machinery [450].

In the winery, advanced fermentation technologies, including temperature-controlled fermentation tanks, ensure consistent wine quality. Innovations such as micro-oxygenation and the use of controlled yeast strains can enhance flavour profiles and improve wine stability [450]. These technological advancements help mitigate the variability often encountered by small-scale winemakers, enabling them to produce higher-quality wines consistently [450].

Sustainability is increasingly vital in winemaking, yet implementing eco-friendly practices can be challenging due to financial constraints. Innovations in sustainable agriculture, such as solar energy systems and water-saving irrigation technologies, can significantly reduce the environmental footprint of vineyards [450]. For example, drip irrigation systems equipped with real-time monitoring can optimize water usage while maintaining vine health [450]. Furthermore, organic farming techniques, supported by tools like soil health monitors, promote long-term sustainability in vineyard management [450].

In the winery, waste-to-value innovations, such as converting grape pomace into biofuel or compost, not only reduce waste but also create additional revenue streams [450]. The adoption of lightweight and recyclable packaging aligns with consumer demand for sustainable products, further enhancing the marketability of wines [450].

Market access remains a significant hurdle for many winemakers, especially small-scale producers lacking the resources of larger competitors. Digital tools and social media platforms provide innovative avenues for connecting with consumers and expanding market reach. E-commerce platforms enable wineries to sell directly to consumers, bypassing traditional distribution channels [433, 451]. Social media platforms like Instagram and Facebook serve as effective marketing tools, allowing winemakers to share their stories and engage with customers [432, 433].

Moreover, the integration of augmented reality (AR) labels and QR codes on wine bottles enhances the consumer experience by providing interactive content, such as vineyard tours and pairing suggestions [433]. These innovations not only educate consumers but also foster brand loyalty by creating memorable connections with the product [433].

Maintaining consistent wine quality across vintages is a significant challenge for winemakers. Innovations in enology, such as advanced testing tools and microbial innovations, can help achieve this goal. Tools like spectrometers and chromatography systems enable precise analysis of grape and wine composition, ensuring that wines meet desired quality standards [450]. Additionally, the use of specific yeast and bacteria strains allows for controlled fermentations that enhance flavour profiles while minimizing spoilage risks [450].

Furthermore, alternatives to traditional oak barrels, such as oak chips or staves, provide cost-effective options for imparting oak flavours without the financial burden of purchasing expensive barrels [450]. These scientific advancements contribute to the overall quality and consistency of wines produced by small-scale winemakers.

Navigating complex regulations and obtaining certifications can be daunting for winemakers. Digital tools, such as compliance management software, streamline the process of tracking and meeting regulatory requirements [450]. Online platforms also provide access to certification programs and training resources, facilitating the achievement of organic or sustainability certifications [450]. Collaborations with industry associations can offer small-scale producers collective bargaining power and shared resources, aiding in navigating regulatory frameworks [450].

Educating consumers about wine and its production process is essential for building brand loyalty and driving sales. Innovative approaches, such as virtual wine tastings and interactive wine education apps, enable winemakers to share their expertise and connect with customers regardless of location [450, 451]. Experiential marketing strategies, including vineyard tours and harvest events, immerse consumers in the winemaking process, creating lasting impressions [450, 451].

By leveraging storytelling and digital tools, small-scale winemakers can differentiate their brands and build deeper connections with their audience, ultimately enhancing their market presence [450, 451].

Future Trends in Sustainable Viticulture

Sustainable viticulture is increasingly recognized as a vital component of the wine industry, driven by the need to address environmental challenges, technological advancements, and consumer preferences for eco-friendly products. The future trends in sustainable viticulture encompass several key areas, including climate-resilient practices, regenerative agriculture, precision viticulture, circular economy approaches, renewable energy integration, advanced pest management, consumer-driven sustainability, water conservation innovations, social and economic sustainability, and collaborative efforts among stakeholders.

As climate change continues to impact vineyard conditions, sustainable viticulture is adopting practices that enhance resilience to extreme weather events, droughts, and temperature fluctuations. The use of heat- and drought-tolerant grape varieties, including hybrids and indigenous species, is becoming more prevalent, as these varieties require fewer inputs and demonstrate greater resilience to changing climates [452, 453]. Advanced irrigation technologies, such as precision drip systems and soil moisture sensors, are being implemented to optimize water usage and minimize waste [452]. Additionally, climate modelling research is aiding growers in anticipating weather patterns, allowing for proactive vineyard management [453]. Techniques such as shade nets and reflective mulches are also being refined to protect vines from heat stress, ensuring consistent yields and quality [454].

Regenerative viticulture is gaining traction as a holistic approach that goes beyond organic and biodynamic practices. This methodology emphasizes restoring soil health, enhancing biodiversity, and capturing carbon to combat climate change [455]. Practices such as cover cropping, minimal tillage, and composting are integral to regenerative vineyards, fostering a healthy ecosystem that supports vine growth [455]. Furthermore, carbon sequestration through improved soil management and agroforestry techniques is becoming a focal point for sustainability initiatives, with vineyards incorporating native vegetation and hedgerows to promote biodiversity while sequestering carbon [455].

The integration of precision agriculture tools is revolutionizing sustainable viticulture. Technologies such as drones, satellite imaging, and sensors provide real-time data on vineyard health, enabling growers to monitor vine stress, disease outbreaks, and soil conditions with high accuracy [452]. This data-driven approach allows for targeted interventions, reducing the use of water, fertilizers, and pesticides [452]. Additionally, robotic technology is emerging in viticulture, with autonomous tractors and drones performing tasks such as pruning, spraying, and harvesting, which enhances efficiency and reduces labour costs [452].

The wine industry is increasingly adopting circular economy principles to minimize waste and maximize resource efficiency. By repurposing grape pomace, stems, and seeds into value-added products such as bioenergy and natural fertilizers, wineries can significantly reduce their environmental footprint [456, 457]. Moreover, the treatment and reuse of winery wastewater for irrigation are becoming standard practices, thereby conserving water and preventing pollution [456]. The adoption of recyclable and biodegradable packaging materials, including lightweight glass bottles and plant-based plastics, is also gaining momentum, catering to environmentally conscious consumers [412].

With rising energy costs and a growing emphasis on sustainability, vineyards are increasingly turning to renewable energy solutions. The installation of solar panels, wind turbines, and geothermal systems is becoming common to power vineyard operations, from irrigation systems to winemaking facilities [456]. Energy-efficient technologies, such as LED lighting and energy recovery systems, further contribute to reducing resource consumption [456]. Additionally, wineries are exploring battery storage solutions and microgrid systems to ensure a stable energy supply, particularly in regions with unreliable power infrastructure [456].

Sustainable viticulture is leveraging biological controls and integrated pest management (IPM) to minimize chemical pesticide use. The deployment of beneficial insects, pheromone traps, and natural predators is becoming more common, while disease-resistant grape varieties are being developed through traditional breeding and genetic research [458]. Digital tools, including artificial intelligence and machine learning, are also being utilized to predict pest outbreaks and disease spread, allowing for precise and timely interventions that reduce environmental impact [458].

The growing consumer demand for sustainably produced wines is influencing the industry to adopt transparent practices. Certifications such as organic, biodynamic, and carbon-neutral labels are becoming essential for market differentiation [456]. Blockchain technology is emerging as a tool for providing traceability, enabling consumers to verify a wine's production practices and origin [456]. Furthermore, sustainable storytelling, where wineries communicate their environmental and social impact initiatives, is gaining popularity, helping to build brand loyalty among eco-conscious consumers [456].

Water scarcity is a pressing concern in many wine-producing regions, prompting the adoption of advanced water management strategies. Techniques such as rainwater harvesting and wastewater recycling are becoming integral to sustainable viticulture [456]. Innovations like precision irrigation and soil moisture monitoring ensure efficient water use without compromising vine health [456]. Some vineyards are also experimenting with dry farming techniques, relying solely on natural rainfall, which not only conserves water but also enhances the terroir-driven characteristics of the wine [456].

Future trends in sustainable viticulture also emphasize social and economic dimensions. Fair wages, safe working conditions, and community engagement are becoming priorities for vineyards [456]. Programs that support local economies, such as farm-to-table initiatives and agritourism, are enhancing the social impact of viticulture [456]. Economic sustainability is being addressed through diversification, with many small-scale vineyards incorporating secondary revenue streams, such as eco-tourism and wine education, to enhance financial resilience [456].

The future of sustainable viticulture relies on collaboration among industry stakeholders, researchers, and policymakers. Knowledge-sharing platforms and collaborative research initiatives are helping vineyards adopt best practices and innovate effectively [456]. Regional sustainability programs, such as Sustainable Winegrowing New Zealand and California Sustainable Winegrowing Alliance, are fostering collective progress towards sustainability goals [456].

Vineyard and Wine Making Health and Safety

Health and safety in vineyard and winemaking operations are paramount for protecting workers, visitors, and the environment. The unique challenges posed by these industries necessitate a comprehensive approach to risk management, particularly in areas such as equipment hazards, chemical exposure, and confined spaces.

Vineyards present various hazards that can lead to injuries or health issues among workers. Equipment hazards are significant, as the use of tractors and mechanized tools can result in accidents if not properly managed. Regular maintenance and training on the safe operation of equipment are essential to mitigate these risks [459, 460]. Furthermore, environmental conditions such as extreme temperatures and UV radiation pose additional threats. Studies have shown that vineyard workers are at risk for heat-related illnesses and skin cancer due to prolonged sun exposure [461]. Implementing protective measures, such as providing shade and hydration stations, is crucial for worker safety [461].

Chemical exposure is another critical concern in vineyard operations. The use of pesticides and herbicides can lead to respiratory issues and skin irritation if not handled correctly. Integrated pest management (IPM) strategies can help reduce reliance on harmful chemicals, while personal protective equipment (PPE) such as gloves and masks is vital for safe handling [459, 462].

The winemaking process introduces additional hazards, particularly in confined spaces such as fermentation tanks. These areas can accumulate hazardous gases like carbon dioxide, leading to asphyxiation risks if proper safety protocols are not followed [460, 463]. Implementing confined space entry protocols and ensuring adequate ventilation are essential strategies for preventing accidents in these environments [464].

Chemical handling in winemaking also poses risks. The use of sulphur dioxide and other additives requires careful management to prevent respiratory problems and chemical burns. Training staff in safe handling procedures and maintaining material safety data sheets (MSDS) are critical components of a comprehensive safety program [462, 465]. Moreover, mechanical hazards from equipment such as crushers and bottling machines necessitate regular safety checks and the use of machine guards to prevent injuries [459, 463].

Sulphur dioxide (SO_2) is a preservative in winemaking, prized for its antimicrobial and antioxidant properties. However, handling SO_2 requires strict safety protocols due to the potential health risks it poses. Exposure to sulphur dioxide can lead to respiratory irritation, skin and eye discomfort, and, in severe cases, significant health complications. Ensuring robust safety measures protects both the workers handling the chemical and the integrity of the winemaking process.

Sulphur dioxide can be used in winemaking in various forms, including as a gas, liquid, or in compounds like potassium or sodium metabisulfite. Its primary hazards include respiratory irritation, as even low concentrations can cause coughing and throat discomfort. Contact with liquid SO_2 or its solutions may result in burns and skin irritation, while high levels of inhalation exposure can lead to serious respiratory conditions, including lung damage. Understanding these risks is critical for implementing effective protective measures.

Proper ventilation is paramount when using SO_2 in winemaking. Workspaces should have exhaust systems to capture fumes at their source and air monitoring systems to detect SO_2 concentrations. These systems alert workers if levels approach hazardous thresholds, ensuring timely action. Equipment used to handle sulphur dioxide, such as tanks, gas cylinders, valves, and hoses, must be inspected and maintained regularly to prevent leaks and ensure safe operation.

Personal protective equipment (PPE) is another cornerstone of sulphur dioxide safety. Workers handling SO_2 must wear certified respirators with SO_2-specific cartridges to protect against inhalation. Goggles or face shields provide essential eye protection, while chemical-resistant gloves and aprons made from materials like neoprene or nitrile safeguard against burns and skin contact. Employers must ensure PPE is properly fitted, stored, and regularly inspected for wear.

The storage of sulphur dioxide also demands careful attention. Containers holding SO_2 or its compounds should be clearly labelled with hazard warnings and handling instructions. The chemical must be stored in cool, dry conditions away from direct sunlight and heat sources, and separate from incompatible substances like strong acids to prevent dangerous reactions.

Training and emergency preparedness are crucial to safe SO_2 use. Workers should be trained to understand Material Safety Data Sheets (MSDS) for sulphur dioxide, as well as emergency procedures for spills, leaks, or accidental exposure. First aid training for treating burns, eye irritation, or respiratory distress is essential. Emergency equipment, such as eyewash stations, safety showers, and spill kits, must be accessible in all areas where SO_2 is used.

Minimizing sulphur dioxide use where possible is a proactive approach to safety. Precision dosing equipment ensures only the minimum effective amount of SO_2 is applied. Alternative preservation methods, such as cold stabilization, sterile filtration, or the use of inert gases like nitrogen, can reduce reliance on SO_2. Additionally, producing low-sulphur wines caters to consumers sensitive to sulphites and aligns with safer operational practices.

Compliance with local regulations governing the use of sulphur dioxide is essential. Detailed records of SO_2 purchases, usage, worker training, and air quality monitoring must be maintained. These records demonstrate adherence to safety standards and can be crucial during audits or inspections.

A robust focus on worker safety is integral to vineyard and winemaking operations. Comprehensive training programs are essential to ensure that workers understand the risks associated with their tasks and the proper use of equipment and chemicals [466]. Topics should include emergency response, first aid, and safe lifting techniques to prevent musculoskeletal injuries [466]. Providing appropriate PPE tailored to specific tasks further enhances worker safety [459].

Health monitoring is also crucial, particularly for workers exposed to hazardous substances or strenuous tasks. Regular health checks can help identify potential issues early, allowing for timely interventions [466].

Wineries often host tours and events, necessitating measures to ensure visitor safety. Clear signage indicating restricted areas and potential hazards is essential for preventing accidents [459]. Emergency preparedness, including readily available first aid kits and trained staff, is vital for addressing unforeseen incidents [459]. Ensuring that facilities are accessible to individuals with disabilities also enhances safety and inclusivity for all guests [459].

Compliance with local health and safety regulations is mandatory for vineyards and wineries. This includes adhering to labour laws, chemical storage guidelines, and workplace safety standards. Maintaining detailed records of safety training, equipment maintenance, and incident reports is crucial for ensuring accountability and supporting regulatory compliance [459, 466].

Creating a culture of safety requires commitment from leadership and active involvement from workers. Encouraging open communication about hazards and rewarding safe practices fosters an environment where health and safety are prioritized [466]. Regular safety audits and continuous improvement initiatives can further enhance the safety culture within the organization [466].

Sustainable practices often align with health and safety goals. For instance, transitioning to organic farming reduces chemical exposure, while renewable energy systems can minimize mechanical risks associated with traditional power sources [459, 462].

Small-scale sustainable winemaking, while environmentally conscious and less resource-intensive than large-scale operations, presents unique hazards that require careful management. These risks stem from the inherent processes of winemaking, the use of chemicals, and the manual nature of tasks involved. Addressing these challenges is essential to ensure the safety of workers, maintain product quality, and protect the environment.

Chemical hazards are a significant concern in small-scale winemaking due to the use of substances like sulphur dioxide, potassium metabisulfite, and cleaning agents. Sulphur dioxide, a common preservative, can cause respiratory irritation, coughing, and breathing difficulties if inhaled. Liquid sulphur dioxide or concentrated solutions pose additional risks, such as skin burns and eye irritation. Cleaning agents, often strong alkaline or acidic solutions like caustic soda or citric acid, can cause chemical burns and respiratory issues if not handled with adequate protective equipment. Even in sustainable practices where pesticide use is minimized, residual chemicals from vineyard operations can pose risks during grape handling and processing, requiring meticulous safety protocols.

Physical hazards are another concern, as winemaking involves a range of labour-intensive activities. Manual grape harvesting exposes workers to potential injuries like cuts, punctures, and repetitive strain injuries, particularly on steep slopes or uneven terrain where slips, trips, and falls are common. Machinery such as crushers, destemmers, and presses pose risks of entanglement, crushing, or cutting injuries, especially if improperly maintained or operated. Additionally, handling heavy grape bins, barrels, or wine cases without proper lifting techniques can lead to musculoskeletal injuries, exacerbating physical strain.

The fermentation process introduces microbiological hazards, primarily through the active microbial activity involved. Carbon dioxide (CO_2) generated during fermentation can displace oxygen in confined spaces, creating a risk of asphyxiation in poorly ventilated areas like cellars or tanks. Certain strains of yeast and bacteria, though generally non-pathogenic, can cause respiratory or skin irritation, particularly for workers with sensitivities or allergies.

Ergonomic and repetitive motion hazards are prevalent in small-scale winemaking due to the manual nature of tasks like corking bottles, labelling, and sorting grapes. These repetitive actions can lead to strain injuries or musculoskeletal disorders, especially during peak harvest and bottling seasons when prolonged work hours contribute to fatigue. Poorly designed workspaces or equipment further exacerbate these issues, leading to long-term health complications.

Environmental and fire hazards also pose risks despite the sustainable focus of small-scale operations. Alcohol vapor and organic residues in confined spaces increase the risk of fire, particularly near open flames or malfunctioning electrical equipment. Improper handling of organic waste, such as grape pomace or wastewater, can result in environmental pollution or pest infestations. Additionally, improper storage of chemicals or wines in spaces with fluctuating temperatures or poor ventilation can lead to leaks, spoilage, and associated safety risks.

Biological hazards remain a concern in sustainable winemaking, as limited pesticide use can make vineyards more vulnerable to pests such as insects, rodents, or mould. These pests not only pose health risks to workers but also impact the quality of the final product.

Machinery used in processes like crushing, destemming, and bottling introduces noise and vibration hazards. Prolonged exposure to high noise levels or vibrations can lead to hearing loss or discomfort over time if not adequately managed through protective measures or equipment maintenance.

Appendices

Appendix A
Small-Scale Winemaking Glossary

Aging: The process of storing wine in barrels, tanks, or bottles to allow it to develop its flavour, aroma, and complexity over time.

Biodynamic Winemaking: A method of winemaking based on the principles of biodynamic farming, focusing on holistic and ecological balance, often incorporating lunar and astrological cycles.

Bladder Press: A pressing device that uses an inflatable bladder to apply gentle, even pressure to grape pomace, extracting juice or wine with minimal tannin extraction.

Brix (°Bx): A measurement of sugar content in grape juice, used to estimate potential alcohol content in the finished wine.

Carbonic Maceration: A winemaking technique where whole grapes ferment in a carbon dioxide-rich environment before being crushed, often used for lighter, fruitier red wines.

Clarification: The process of removing suspended particles from wine, improving its clarity and stability, often through methods such as fining, filtration, or cold stabilization.

Crushing: Breaking the grape skins to release juice before fermentation, typically done with a mechanical crusher or by hand for small batches.

Decanting: The act of pouring wine from its bottle into another container to separate it from sediment or to aerate it for better flavour and aroma.

Destemming: Removing the stems from grape clusters before fermentation, common in red wine production to reduce tannin and vegetal flavours.

Fermentation: The biochemical process where yeast converts sugars in grape juice into alcohol and carbon dioxide, creating wine.

Fining: A clarification process where substances like bentonite, egg whites, or gelatin are added to wine to bind and remove unwanted particles or flavours.

Geographical Indication (GI): A certification that links a wine to its specific region of origin, ensuring it meets the region's established production standards.

Lees: The sediment of dead yeast cells and grape solids that settles at the bottom of fermentation vessels; often stirred in (bâtonnage) for added texture and flavour in some wines.

Malolactic Fermentation (MLF): A secondary fermentation process where malic acid is converted into lactic acid, softening the wine and often adding creamy or buttery notes.

Maceration: The process of soaking grape skins, seeds, and stems in the juice during fermentation to extract colour, flavour, and tannins.

Natural Wine: Wine made with minimal intervention, typically avoiding additives like sulphur dioxide and utilizing wild fermentation.

Organic Winemaking: A method of winemaking using organically grown grapes and adhering to organic certification standards, often with limited use of additives.

Pigeage: The French term for "punching down" the grape skins during fermentation to enhance extraction and prevent the cap from drying out.

Pomace: The solid remains of grapes after pressing, including skins, seeds, and stems, which can be composted or distilled.

Racking: Transferring wine from one container to another to separate it from sediment (lees) and clarify it.

Sulphites (SO₂): Preservatives added to wine to prevent oxidation and microbial spoilage; natural sulphites are also produced during fermentation.

Tannin: Naturally occurring compounds in grape skins, seeds, and stems that contribute to the wine's astringency, structure, and aging potential.

Terroir: The unique combination of soil, climate, and geographical factors that influence the character and flavour profile of wine from a specific region.

Vintner: A person or company involved in the production of wine.

Vinification: The process of converting grape juice into wine through fermentation and other winemaking techniques.

Wild Fermentation: A fermentation process that relies on naturally occurring yeast from the grapes or winery environment rather than cultured yeast strains.

Yeast: Microorganisms that ferment grape sugars into alcohol and carbon dioxide, essential for winemaking.

Yield: The amount of wine produced from a vineyard or batch of grapes, often measured in tons per acre or hectoliters per hectare.

Appendix B

Templates for Vineyard Planning and Management

Effective vineyard planning and management require the use of structured templates to ensure thorough planning, efficient resource use, and consistent monitoring. These templates help streamline processes, maintain records, and align activities with business goals, environmental practices, and legal requirements. Following are the essential templates that small-scale vineyard operators can use for comprehensive planning and management:

Vineyard Site Assessment Template

Purpose

Evaluate the suitability of a site for vine planting to ensure optimal grapevine growth, yield quality, and sustainable operations.

Key Sections

1. Soil Type and Quality

- **Soil Texture**: [Insert details on soil texture, such as sandy, loamy, or clay].

- **pH Level**: [Insert soil pH test results].

- **Nutrient Content**: [Insert details on nutrient levels, including nitrogen, phosphorus, potassium].

- **Drainage Capacity**: [Describe drainage conditions, such as well-drained, waterlogged, etc.].

- **Soil Organic Matter**: [Insert percentage of organic matter].

- **Recommendations for Improvement**: [Insert suggested soil amendments or treatments].

2. Climate Data

- **Annual Temperature Range**: [Insert average high and low temperatures].

- **Rainfall Levels**: [Insert annual and seasonal rainfall data].

- **Frost Risks**: [Describe frost occurrence, including timing and intensity].

- **Sunlight Hours**: [Insert annual average hours of sunlight].

3. Water Availability and Irrigation Potential

- **Water Sources**: [Describe available water sources, such as wells, reservoirs, or rivers].

- **Water Quality**: [Insert results from water quality tests, focusing on salinity and pH].

- **Irrigation Feasibility**: [Describe the suitability of the site for irrigation systems, such as drip or overhead irrigation].

4. Topography

- **Slope**: [Insert degree of slope and its orientation].

- **Aspect**: [Describe the site's exposure to sunlight, such as north-facing or south-facing].

- **Elevation**: [Insert site elevation above sea level].

- **Erosion Risks**: [Describe potential erosion risks based on slope and soil conditions].

5. Accessibility and Proximity to Markets or Processing Facilities

- **Road Access**: [Describe the quality and availability of roads leading to the site].

- **Distance to Markets**: [Insert distance to the nearest local, regional, or export markets].

- **Proximity to Processing Facilities**: [Insert distance to wineries or grape processing units].

- **Transport Challenges**: [Describe potential challenges, such as remote location or seasonal access issues].

6. Environmental Impact and Biodiversity Considerations

- **Existing Vegetation**: [Describe the current plant and tree cover].

- **Biodiversity**: [Insert observations on local flora and fauna, including protected species].

- **Impact on Surrounding Areas**: [Describe potential effects on nearby ecosystems].

- **Sustainability Practices**: [List proposed measures, such as cover cropping, wildlife corridors, or organic farming].

Conclusion and Recommendations

- **Overall Suitability**: [Summarize findings on site suitability for viticulture].

- **Next Steps**: [List suggested actions, such as soil improvement, erosion control, or frost protection measures].

- **Long-Term Sustainability Considerations**: [Provide suggestions for ensuring sustainable vineyard operations].

Vineyard Layout and Design Template

Purpose

Plan vineyard block layouts to optimize space utilization, improve vineyard productivity, and ensure sustainable management practices.

Key Sections

1. Row Orientation and Spacing

- **Row Orientation**: [Describe orientation, e.g., north-south, east-west, and reasoning for choice based on sunlight and airflow].

- **Row Spacing**: [Insert spacing between rows, e.g., 2.5m, and justification based on equipment size and vine variety].

- **Vine Spacing Within Rows**: [Insert distance between vines, e.g., 1.2m, and its impact on vine growth and yield].

- **Access Paths**: [Indicate location and width of paths for equipment and workers].

2. Trellis System Design

- **Trellis Type**: [Describe the chosen trellis system, e.g., Vertical Shoot Positioning (VSP), Geneva Double Curtain, etc.].

- **Materials**: [List materials for trellis construction, such as wood, metal, or composite posts].

- **Wire Configuration**: [Detail the number of wires, spacing, and tension requirements].

- **Future Adaptability**: [Describe provisions for system adjustments as the vineyard matures].

3. Vine Variety Allocation by Block

- **Block Designation**: [List block names or numbers].

- **Variety Allocation**: [Specify vine varieties planted in each block].

- **Rationale for Allocation**: [Explain decisions based on soil type, microclimate, or market demand].

- **Yield Goals**: [Insert expected yield per variety or block].

4. Drainage and Erosion Control Features

- **Drainage Plan**: [Describe drainage infrastructure, such as trenches, tiles, or contour drains].

- **Erosion Control Measures**: [Insert features like cover crops, terraces, or berms to minimize soil loss].

- **Water Flow Mapping**: [Include details of water flow across the vineyard and areas at risk].

- **Sustainable Practices**: [Detail methods like planting grass strips or mulching to enhance soil stability].

5. Buffer Zones for Biodiversity or Environmental Protection

- **Buffer Zone Location**: [Mark the areas designated as buffer zones, e.g., at vineyard edges or along water bodies].

- **Purpose**: [Describe the role of the buffer zones, such as protecting waterways or fostering wildlife habitats].

- **Vegetation Plan**: [List plants or trees to be used in buffer zones to enhance biodiversity].

- **Compliance**: [Insert any regulatory requirements for buffer zones].

Supplementary Features

- **Irrigation System Layout**: [Map irrigation lines and equipment for efficient water use].

- **Windbreaks**: [Describe features such as tree rows or fences to protect vines from strong winds].

- **Energy Infrastructure**: [Plan for renewable energy sources like solar panels or wind turbines if applicable].

Conclusion and Recommendations

- **Final Layout Summary**: [Provide an overview of the vineyard layout and its alignment with production and sustainability goals].

- **Next Steps**: [List actions required to finalize and implement the vineyard layout].

- **Long-Term Adaptation Plan**: [Highlight strategies for future vineyard expansion or modifications].

Planting Schedule and Budget Template

Purpose

Organize and track planting activities, resources, and costs to ensure efficient vineyard establishment and financial planning.

Key Sections

1. Planting Timeline

- **Start Date**: [Insert planned start date for planting activities].
- **Key Milestones**:
    - Soil preparation completion: [Date]
    - Trellis installation: [Date]
    - Vine delivery: [Date]
    - Planting commencement: [Date]
    - Initial irrigation setup: [Date]
    - Post-planting review and adjustments: [Date]
- **End Date**: [Insert expected completion date].

2. Varietal Selection and Sourcing Details

- **Selected Varieties**:
    - Variety 1: [Name, e.g., Cabernet Sauvignon]
    - Variety 2: [Name, e.g., Chardonnay]
    - Variety 3: [Name, e.g., Pinot Noir]

- **Source of Vines**:
    - ○ Supplier Name: [Insert name of supplier].
    - ○ Contact Information: [Insert supplier contact details].
    - ○ Certification: [Specify if organic, disease-free, or certified stock].
- **Quantity Ordered**: [Insert the total number of vines per variety].
- **Delivery Date**: [Insert expected delivery date].

3. Labour and Equipment Needs

- **Labor Requirements**:
    - ○ Total workers needed: [Insert number].
    - ○ Skilled labour: [Insert details, e.g., trellis installation, planting supervision].
    - ○ General labour: [Insert details, e.g., vine placement, staking].
    - ○ Estimated labour hours: [Insert total hours].
- **Equipment Requirements**:
    - ○ Equipment Type: [List equipment, e.g., tractor, augers, irrigation tools].
    - ○ Quantity: [Insert required number].
    - ○ Rental or Ownership: [Specify whether renting or using owned equipment].
- **Availability Dates**: [Insert planned dates for equipment use].

4. Cost Estimates

- **Vines**:

- o Variety 1: [Name] | Quantity: [Number] | Cost per vine: [Amount] | Total: [Amount]

- o Variety 2: [Name] | Quantity: [Number] | Cost per vine: [Amount] | Total: [Amount]

- **Stakes and Trellises**:

  - o Stakes: [Quantity] | Cost per stake: [Amount] | Total: [Amount]

  - o Trellis materials: [List materials] | Cost: [Amount]

- **Labor Costs**:

  - o Skilled labour: [Hourly rate] | Hours: [Total] | Total: [Amount]

  - o General labour: [Hourly rate] | Hours: [Total] | Total: [Amount]

- **Additional Costs**:

  - o Equipment rental: [Itemized costs].

  - o Transportation: [Insert costs for delivery of vines and equipment].

  - o Miscellaneous: [Include unexpected costs].

5. Total Budget Overview

- **Total Estimated Costs**: [Insert grand total of all costs].

- **Contingency Fund**: [Insert percentage or amount for unforeseen expenses].

- **Final Budget Allocation**: [Insert total including contingency].

Conclusion and Notes

- **Project Summary**: [Provide a brief overview of the planting schedule and budget].

- **Key Considerations**: [Highlight potential risks, such as delayed vine delivery or labour shortages].

- **Next Steps**: [List any additional actions required before planting begins].

Vineyard Maintenance Calendar Template

Purpose

Organize and schedule routine tasks to maintain vineyard health, productivity, and sustainability throughout the year.

Key Sections

1. Pruning and Training Schedules

- **Winter Dormant Pruning**:
    - Target Dates: [Insert date range].
    - Objectives: Remove dead wood, shape vines for optimal growth.
    - Tools Needed: [Insert tools, e.g., pruning shears, loppers].
- **Summer Training**:
    - Target Dates: [Insert date range].
    - Objectives: Position shoots, remove suckers, optimize airflow and sunlight.
    - Methods: [Describe training system, e.g., Vertical Shoot Positioning (VSP)].

2. Irrigation and Water Management

- **Seasonal Watering Schedule**:
    - Spring: [Frequency and volume].
    - Summer: [Frequency and volume].
    - Fall: [Frequency and volume].

- **Monitoring and Adjustments**:
  - o Tools: [Insert tools, e.g., tensiometers, soil moisture probes].
  - o Notes: Adjust irrigation based on rainfall and vine growth stage.

3. Soil Amendments and Fertilization Timelines

- **Soil Testing**:
  - o Target Dates: [Insert dates for testing].
  - o Parameters: pH, nutrient levels, organic matter.
- **Fertilization Plan**:
  - o Spring Application: [Type of fertilizer and application method].
  - o Mid-Season Application: [Type of fertilizer and application method].
- **Composting**:
  - o Materials: [Insert materials, e.g., grape pomace, organic waste].
  - o Frequency: [Insert compost application schedule].

4. Pest and Disease Control Measures

- **Monitoring Schedule**:
  - o Frequency: Weekly/biweekly inspections.
  - o Tools: [Insert tools, e.g., magnifying glasses, pest traps].
- **Integrated Pest Management (IPM) Strategies**:
  - o Preventative Measures: [Insert, e.g., cover crops, natural predators].
  - o Control Measures: [Insert, e.g., organic sprays, biological agents].
- **Disease Control**:

o   Spring: [Insert measures, e.g., fungicide application].

o   Summer: [Insert measures, e.g., canopy thinning for airflow].

5. Seasonal Canopy Management

- **Spring**:

    o   Task: Shoot thinning and positioning.

    o   Objective: Promote balanced growth and airflow.

- **Summer**:

    o   Task: Leaf removal in fruit zones.

    o   Objective: Enhance sunlight exposure for fruit ripening.

- **Fall**:

    o   Task: Final canopy trimming.

    o   Objective: Prepare vines for harvest and reduce disease risks.

6. Notes and Adjustments

- **Weather Considerations**: [Insert potential impacts and adjustments].

- **Equipment Maintenance**: [Insert schedules for tool sharpening, equipment repair].

- **Special Instructions**: [Insert details for unique vineyard needs or events].

Annual Maintenance Overview

- **Yearly Goals**: [Summarize objectives, e.g., yield targets, quality improvements].

- **Key Challenges**: [Identify potential risks or areas of concern].

- **Action Plan**: [Outline steps to address challenges].

Pest and Disease Management Template

Purpose

To monitor, track, and manage vineyard pests and diseases effectively, ensuring healthy vine growth and optimal grape production while minimizing environmental impact.

Key Sections

1. Common Vineyard Pests and Diseases

- **Pests**:

    o [Insert pests, e.g., grapevine moth, spider mites, leafhoppers].

    o Description: [Briefly describe the pest and its lifecycle].

    o Damage: [Explain the damage caused, e.g., defoliation, fruit damage].

- **Diseases**:

    o [Insert diseases, e.g., powdery mildew, botrytis bunch rot, downy mildew].

    o Cause: [Insert, e.g., fungal, bacterial, or viral].

    o Symptoms: [Describe visual signs, e.g., leaf discoloration, fruit rot].

2. Monitoring Schedules and Thresholds for Intervention

- **Pest Monitoring**:

    o Frequency: [Insert, e.g., weekly inspections].

    o Methods: [Insert, e.g., pheromone traps, visual scouting].

    o Thresholds: [Define intervention levels, e.g., 5% infestation rate].

- **Disease Monitoring**:

  - Frequency: [Insert, e.g., biweekly during high-risk seasons].

  - Methods: [Insert, e.g., leaf sampling, moisture level checks].

  - Thresholds: [Insert, e.g., visible symptoms on 10% of leaves].

- **Weather Data Tracking**:

  - Parameters: [Insert, e.g., humidity, rainfall, temperature].

  - Tools: [Insert, e.g., weather stations, moisture sensors].

3. Treatment Records

- **Date of Treatment**: [Insert date].

- **Target Pest or Disease**: [Insert name].

- **Method of Control**:

  - Chemical: [Insert product name, active ingredient, dosage].

  - Biological: [Insert, e.g., beneficial insects, microbial sprays].

  - Cultural Practices: [Insert, e.g., canopy management, crop rotation].

- **Application Details**:

  - Equipment Used: [Insert, e.g., sprayers, drones].

  - Area Treated: [Insert block or vineyard section].

  - Weather Conditions: [Insert temperature, wind, humidity].

- **Follow-Up Results**:

  - Effectiveness: [Insert, e.g., % reduction in pest population].

  - Notes: [Record observations, successes, or challenges].

4. Integrated Pest Management (IPM) Strategy Tracking

- **Preventative Measures**:
    - o [Insert, e.g., cover crops, pruning for airflow, resistant grape varieties].
    - o Implementation Date: [Insert dates].
- **Biological Controls**:
    - o [Insert, e.g., release of predatory insects, microbial solutions].
    - o Frequency: [Insert intervals or timing].
- **Chemical Controls**:
    - o Threshold for Use: [Define acceptable levels for chemical application].
    - o Products Allowed: [List approved pesticides/herbicides].
- **Cultural Practices**:
    - o [Insert, e.g., canopy management, sanitation practices].
    - o Description: [Explain specific techniques used].
- **Evaluation and Adjustments**:
    - o Monitoring Outcomes: [Summarize results of interventions].
    - o Strategy Updates: [Detail changes made to improve IPM effectiveness].

5. Annual Summary and Insights

- **Pest and Disease Trends**: [Identify recurring issues, seasonal patterns].
- **Effectiveness of Treatments**: [Evaluate success rates of various interventions].
- **Environmental Impact**: [Assess progress in reducing chemical use and supporting biodiversity].

- **Future Goals**: [Outline objectives for improved pest and disease management].

Vineyard Health and Soil Monitoring Template

Purpose

To systematically monitor and document soil conditions and vine health to support sustainable vineyard management, optimize grape quality, and ensure long-term vineyard productivity.

Key Sections

1. Soil Test Results

- **Sampling Date**: [Insert date].

- **Location**: [Specify vineyard block or section].

- **Test Parameters**:

  - pH: [Insert result, e.g., 6.5].

  - Nutrient Levels:

    - Nitrogen (N): [Insert result].

    - Phosphorus (P): [Insert result].

    - Potassium (K): [Insert result].

    - Additional Nutrients: [Insert results for calcium, magnesium, etc.].

  - Organic Matter Content: [Insert percentage].

  - Soil Texture and Composition: [Insert, e.g., sandy loam, clay].

  - Drainage Characteristics: [Insert, e.g., good, moderate, poor].

- **Analysis Notes**: [Summarize test findings, highlight deficiencies or imbalances].

2. Vine Health Indicators

- **Assessment Date**: [Insert date].

- **Block or Section Assessed**: [Specify location].

- **Growth Rate**:

  o Shoot Length: [Insert average length, e.g., 20 cm].

  o Rate of Development: [Insert, e.g., on target, delayed].

- **Foliage Condition**:

  o Colour: [Insert, e.g., healthy green, yellowing].

  o Texture: [Insert, e.g., firm, wilted].

  o Presence of Pests or Diseases: [Detail findings, e.g., powdery mildew observed].

- **Fruit Development**:

  o Cluster Formation: [Insert status, e.g., uniform, irregular].

  o Berry Condition: [Insert observations, e.g., firm, underdeveloped].

- **Overall Vine Health Rating**: [Insert, e.g., excellent, fair, poor].

3. Weather Data and Impacts on Growth

- **Monitoring Period**: [Insert start and end dates].

- **Weather Metrics**:

  o Temperature Range: [Insert, e.g., 18°C–30°C].

  o Rainfall: [Insert total, e.g., 25 mm].

  o Frost Events: [Insert dates and severity].

  o Wind Conditions: [Insert, e.g., mild, strong gusts].

- **Impact Analysis**:

- o Effects on Soil: [Insert, e.g., waterlogging, drought stress].
- o Effects on Vines: [Insert, e.g., slowed growth, damaged leaves].

4. Recommendations for Soil Amendments or Adjustments

- **Amendments Needed**:
  - o pH Adjustments: [Insert, e.g., lime for acidic soil].
  - o Nutrient Additions: [Specify, e.g., add 20 kg/ha of potassium sulphate].
  - o Organic Matter: [Insert, e.g., incorporate compost or cover crops].

- **Irrigation Modifications**:
  - o Suggested Changes: [Insert, e.g., increase frequency during dry periods].

- **Erosion or Drainage Improvements**:
  - o Recommended Actions: [Insert, e.g., install contour drains].

- **Pest and Disease Mitigation**:
  - o Strategies: [Insert, e.g., apply organic fungicide, improve airflow].

- **Monitoring Frequency Adjustments**:
  - o Proposed Changes: [Insert, e.g., shift to biweekly soil testing].

5. Annual Summary and Action Plan

- **Observations and Trends**: [Summarize key findings for the year].

- **Effectiveness of Interventions**: [Evaluate success of previous recommendations].

- **Goals for Next Year**: [Outline objectives, e.g., improve organic matter by 5%, reduce pest incidence by 10%].

Irrigation Management Template

Purpose

To optimize water use and improve irrigation efficiency in the vineyard, ensuring the health and productivity of vines while conserving resources and maintaining sustainability.

Key Sections

1. Water Source and Quality Analysis

- **Water Source**: [Specify source, e.g., groundwater, surface water, rainwater collection].

- **Quality Parameters**:

    o pH Level: [Insert value, e.g., 7.0].

    o Salinity: [Insert value, e.g., 0.5 dS/m].

    o Mineral Content: [Specify concentrations of calcium, magnesium, etc.].

    o Contaminants: [Insert findings, e.g., none, moderate].

- **Testing Frequency**: [Insert schedule, e.g., quarterly].

- **Recommendations for Improvement**: [Insert, e.g., install a filtration system].

2. Irrigation System Type and Maintenance Schedule

- **System Type**:

    o [Insert, e.g., drip irrigation, sprinkler, furrow].

- **System Components**:

    o Pipes and Valves: [Insert status, e.g., inspected quarterly].

- o   Emitters: [Insert type and flow rate, e.g., 2 L/hour].

- o   Pumps: [Insert details, e.g., submersible pump, inspected monthly].

- **Maintenance Schedule**:

  - o   Cleaning Emitters: [Insert frequency, e.g., every 3 months].

  - o   Checking for Leaks: [Insert, e.g., monthly visual inspection].

  - o   System Calibration: [Insert, e.g., annually].

- **Notes on Repairs/Upgrades**: [Insert, e.g., replaced faulty emitters in Block A].

3. Seasonal Water Demand for Each Vineyard Block

- **Block Details**:

  - o   Block Name: [Insert, e.g., Block A].

  - o   Vine Variety: [Insert, e.g., Cabernet Sauvignon].

  - o   Vine Age: [Insert, e.g., 5 years].

- **Seasonal Water Requirements**:

  - o   Spring: [Insert volume, e.g., 20 mm/week].

  - o   Summer: [Insert volume, e.g., 50 mm/week].

  - o   Autumn: [Insert volume, e.g., 30 mm/week].

  - o   Winter: [Insert volume, e.g., no irrigation].

- **Adjustments Based on Weather Data**: [Insert, e.g., increased irrigation by 10% during heatwave].

4. Record of Water Usage and Adjustments

- **Date**: [Insert date].

- **Block**: [Insert, e.g., Block A].

- **Volume Applied**: [Insert, e.g., 2,000 liters].

- **Method**: [Insert, e.g., drip irrigation].

- **Reason for Adjustment**: [Insert, e.g., high temperatures, reduced rainfall].

- **Cumulative Monthly Usage**: [Insert, e.g., 8,000 liters for July].

- **Efficiency Observations**: [Insert, e.g., uniform water delivery, minor pooling in Block B].

5. Annual Summary and Recommendations

- **Total Water Used**: [Insert total volume, e.g., 120,000 liters].

- **System Performance**: [Insert, e.g., no major issues, minor clogging in emitters].

- **Adaptations for Next Year**: [Insert, e.g., upgrade filtration, adjust summer irrigation schedule].

- **Sustainability Goals**: [Insert, e.g., reduce water usage by 15%, implement rainwater harvesting].

Harvest Planning Template

Purpose

To efficiently organize harvest logistics while maintaining the quality of grapes, optimizing resources, and ensuring seamless transitions to post-harvest processes.

Key Sections

1. Expected Harvest Dates for Each Varietal

- **Block Name/Number**: [Insert block identifier, e.g., Block A].

- **Varietal**: [Insert, e.g., Chardonnay].

- **Expected Harvest Date**: [Insert date, e.g., September 15].

- **Harvest Window**: [Insert duration, e.g., 3 days].

- **Ripeness Indicators**:

    o   Brix: [Insert target level, e.g., 22°].

    o   Acidity: [Insert, e.g., 6 g/L].

    o   pH: [Insert, e.g., 3.4].

2. Equipment and Labor Needs

- **Labor Requirements**:

    o   Number of Workers: [Insert, e.g., 10 pickers].

    o   Specialized Roles: [Insert, e.g., quality checkers, transport coordinators].

    o   Schedule: [Insert, e.g., 6:00 AM – 2:00 PM].

- **Equipment List**:

- o Harvest Bins: [Insert quantity, e.g., 50 bins].

- o Picking Shears: [Insert quantity, e.g., 20 pairs].

- o Harvest Crates: [Insert, e.g., 30 reusable crates].

- o Machinery (if applicable): [Insert, e.g., 1 mechanical harvester].

- **Maintenance and Preparation**:

  - o Pre-harvest Equipment Inspection: [Insert checklist for functionality].

  - o Repair and Cleaning: [Insert actions, e.g., sharpen shears, sanitize bins].

3. Grape Transport Logistics

- **Transport Method**: [Insert, e.g., flatbed truck, refrigerated truck].

- **Capacity**: [Insert, e.g., 2,000 kg per trip].

- **Transport Timing**: [Insert, e.g., every 2 hours].

- **Distance to Winery**: [Insert, e.g., 15 km].

- **Temperature Control Measures**: [Insert, e.g., shaded bins, refrigerated transport].

- **Contact Information for Drivers**: [Insert names and phone numbers].

4. Quality Assessment Criteria at Harvest

- **Visual Checks**:

  - o Grape Colour: [Insert, e.g., golden yellow for Chardonnay].

  - o Bunch Condition: [Insert, e.g., no shrivelled or mouldy berries].

- **Laboratory Analysis**:

  - o Brix Levels: [Insert acceptable range, e.g., 22°–24°].

- o   Acidity: [Insert, e.g., 5–7 g/L].

- o   pH Levels: [Insert, e.g., 3.3–3.6].

- **Sorting Protocol**:

  - o   In-field Sorting: [Insert, e.g., remove damaged clusters].

  - o   Winery Sorting: [Insert, e.g., manual sorting on the conveyor belt].

5. Post-Harvest Plans

- **Crushing**:

  - o   Location: [Insert, e.g., winery press room].

  - o   Schedule: [Insert, e.g., within 4 hours of harvest].

  - o   Equipment: [Insert, e.g., crusher-destemmer].

- **Storage**:

  - o   Type: [Insert, e.g., stainless steel tanks, oak barrels].

  - o   Conditions: [Insert, e.g., temperature at 12°C].

- **Processing Plans**:

  - o   White Wines: [Insert, e.g., direct pressing after destemming].

  - o   Red Wines: [Insert, e.g., maceration with skins].

  - o   Rosé Wines: [Insert, e.g., short skin contact, then pressing].

- **Cleaning and Maintenance**:

  - o   Post-Harvest Equipment Cleaning: [Insert tasks, e.g., sanitize bins and presses].

  - o   Storage Area Preparation: [Insert, e.g., ensure tanks are sterilized].

Financial Planning and Budget Template

Purpose

To monitor vineyard investments, operational costs, revenue streams, and overall profitability, enabling effective financial management and sustainable growth.

Key Sections

1. Initial Capital Expenditure Breakdown

- **Land Acquisition Costs**:
    - Purchase Price: [Insert amount, e.g., $200,000].
    - Legal Fees: [Insert amount, e.g., $5,000].
    - Surveying and Appraisals: [Insert amount, e.g., $2,000].

- **Infrastructure Development**:
    - Irrigation Systems: [Insert amount, e.g., $10,000].
    - Fencing: [Insert amount, e.g., $3,000].
    - Vineyard Layout and Planting: [Insert amount, e.g., $15,000].

- **Equipment Purchases**:
    - Tractors and Tools: [Insert amount, e.g., $20,000].
    - Fermentation Tanks: [Insert amount, e.g., $12,000].
    - Harvesting Equipment: [Insert amount, e.g., $8,000].

- **Other Initial Investments**:
    - Licenses and Permits: [Insert amount, e.g., $3,000].
    - Initial Vineyard Stock: [Insert amount, e.g., $10,000].

  - Storage Facilities: [Insert amount, e.g., $7,000].

2. Operating Expenses

- **Labor Costs**:

  - Seasonal Labor: [Insert amount, e.g., $5,000/year].

  - Permanent Staff Salaries: [Insert amount, e.g., $25,000/year].

- **Material Costs**:

  - Fertilizers and Soil Amendments: [Insert amount, e.g., $2,000/year].

  - Pesticides and Herbicides: [Insert amount, e.g., $1,500/year].

  - Bottling and Packaging: [Insert amount, e.g., $4,000/year].

- **Maintenance Expenses**:

  - Equipment Maintenance: [Insert amount, e.g., $3,000/year].

  - Infrastructure Repairs: [Insert amount, e.g., $1,000/year].

- **Utilities and Operational Costs**:

  - Water and Irrigation: [Insert amount, e.g., $2,500/year].

  - Energy Costs: [Insert amount, e.g., $3,000/year].

  - Miscellaneous Operational Expenses: [Insert amount, e.g., $1,000/year].

3. Revenue Projections

- **Grape Sales**:

  - Expected Yield (tons): [Insert amount, e.g., 10 tons/year].

  - Price per Ton: [Insert amount, e.g., $2,000/ton].

  - Total Grape Sales Revenue: [Insert amount, e.g., $20,000/year].

- **Wine Production**:

    o Bottles Produced: [Insert amount, e.g., 5,000 bottles/year].

    o Average Price per Bottle: [Insert amount, e.g., $15/bottle].

    o Total Wine Sales Revenue: [Insert amount, e.g., $75,000/year].

- **Additional Revenue Streams**:

    o Tasting Room Sales: [Insert amount, e.g., $10,000/year].

    o Events and Tours: [Insert amount, e.g., $5,000/year].

4. Profitability Analysis and Cash Flow Tracking

- **Profitability Analysis**:

    o Total Revenue: [Insert amount, e.g., $100,000/year].

    o Total Operating Costs: [Insert amount, e.g., $50,000/year].

    o Net Profit: [Insert amount, e.g., $50,000/year].

- **Cash Flow Tracking**:

    o Monthly Revenue: [Insert table or breakdown for monthly cash inflows].

    o Monthly Expenses: [Insert table or breakdown for monthly cash outflows].

    o Net Cash Flow: [Insert, e.g., $5,000 surplus per month].

- **Financial Ratios and Metrics**:

    o Break-Even Analysis: [Insert calculation or chart].

    o Return on Investment (ROI): [Insert, e.g., 25%].

    o Debt-to-Equity Ratio: [Insert, e.g., 0.3].

Sustainability and Certification Tracking Template

Purpose

To document, monitor, and manage sustainable practices and certification requirements within the vineyard and winemaking operations, ensuring alignment with environmental goals and regulatory standards.

Key Sections

1. Water and Energy Conservation Efforts

- **Water Usage**:
    - Total water consumption (monthly/annually): [Insert data].
    - Efficiency measures implemented (e.g., drip irrigation, rainwater harvesting): [Insert details].
    - Monitoring tools used (e.g., soil moisture sensors, water meters): [Insert tools].

- **Energy Usage**:
    - Total energy consumption (monthly/annually): [Insert data].
    - Renewable energy sources (e.g., solar panels, wind turbines): [Insert details].
    - Energy-efficient equipment or practices (e.g., LED lighting, optimized machinery): [Insert details].

2. Organic or Biodynamic Practices

- **Fertilizers and Soil Health**:
    - Organic fertilizers used: [Insert types and amounts].

- o Soil health improvement practices (e.g., cover cropping, composting): [Insert practices].

- **Pest and Disease Management**:

  - o Integrated Pest Management (IPM) strategies: [Insert details].

  - o Biological controls implemented (e.g., beneficial insects, natural predators): [Insert details].

- **Biodynamic Certification Requirements**:

  - o Preparations and applications (e.g., biodynamic sprays, soil enhancers): [Insert details].

  - o Compliance with lunar or celestial planting calendars: [Insert details].

3. Waste Management Strategies

- **Organic Waste**:

  - o Composting practices (e.g., pomace, stems, and leaves): [Insert details].

  - o Reuse initiatives (e.g., grape seed oil production, animal feed): [Insert details].

- **Water Waste**:

  - o Treatment methods for winery wastewater: [Insert methods].

  - o Reuse systems implemented (e.g., irrigation, cleaning): [Insert systems].

- **Packaging Waste**:

  - o Use of recycled or biodegradable materials: [Insert details].

  - o Recycling initiatives for bottles, corks, and packaging: [Insert details].

4. Certification Progress and Compliance Documentation

- **Certification Goals and Timeline**:
    - Target certifications (e.g., Organic, Biodynamic, LEED, Sustainable Winegrowing): [Insert goals].
    - Expected completion dates: [Insert dates].

- **Compliance Activities**:
    - Documentation of required processes (e.g., audit logs, test results): [Insert details].
    - Staff training on certification standards: [Insert training details].

- **Progress Tracking**:
    - Milestones achieved: [Insert achievements and dates].
    - Pending actions: [Insert tasks and timelines].

- **Regulatory Documentation**:
    - Record of inspections and audits: [Insert records].
    - Certification applications submitted: [Insert details and dates].

Notes and Additional Observations

[Insert additional comments or insights related to sustainability practices and certification progress.]

Employee Training and Safety Template

Purpose

To ensure proper training and compliance with safety protocols for all vineyard and winery workers, promoting a safe and efficient working environment.

Key Sections

1. Training Records for Vineyard Workers

- **Employee Details**:
    - Name: [Insert name].
    - Job Title: [Insert title].
    - Start Date: [Insert date].

- **Training Topics Completed**:
    - General Vineyard Operations: [Insert date].
    - Equipment Handling and Maintenance: [Insert date].
    - Chemical Safety and Handling: [Insert date].
    - Personal Protective Equipment (PPE) Usage: [Insert date].

- **Trainer Information**:
    - Name of Trainer: [Insert name].
    - Organization or Certification Body: [Insert details].

- **Renewal or Recertification Dates**:
    - Next Training Date: [Insert date].

2. Safety Protocols for Equipment Use

- **Equipment List and Protocols**:
  - o Name of Equipment: [Insert equipment name].
  - o Safety Instructions: [Insert details].
  - o Maintenance Schedule: [Insert schedule].
- **Authorized Operators**:
  - o Name of Authorized Employee(s): [Insert names].
- **Incident Reporting**:
  - o Procedure for Reporting Malfunctions: [Insert details].
  - o Documentation Required: [Insert forms or steps].

3. Chemical Handling and Personal Protective Equipment (PPE) Requirements

- **Chemical Inventory**:
  - o Name of Chemical: [Insert name].
  - o Safety Data Sheet (SDS) Available: [Yes/No].
- **PPE Requirements for Handling Chemicals**:
  - o Gloves: [Insert type].
  - o Respirators: [Insert type].
  - o Eye Protection: [Insert type].
  - o Protective Clothing: [Insert type].
- **Chemical Handling Training**:
  - o Date of Training: [Insert date].
  - o Trainer: [Insert name].
  - o Refresher Course Due Date: [Insert date].
- **Spill and Exposure Protocols**:

- o Spill Containment Procedure: [Insert steps].

- o First Aid Measures: [Insert steps].

4. Emergency Contact Information and Procedures

- **Emergency Contacts**:

  - o Manager/Supervisor: [Insert name and phone number].

  - o Local Emergency Services: [Insert numbers].

  - o Poison Control Hotline: [Insert number].

- **Emergency Procedures**:

  - o Fire or Evacuation Plan: [Insert steps].

  - o Medical Emergency Response: [Insert steps].

  - o Chemical Exposure Response: [Insert steps].

- **Emergency Equipment**:

  - o Location of First Aid Kits: [Insert locations].

  - o Location of Fire Extinguishers: [Insert locations].

  - o Emergency Shower and Eyewash Stations: [Insert locations].

Notes and Observations

[Insert additional comments or observations related to employee safety and training.]

Vineyard Productivity and Yield Tracking Template

Purpose

To systematically record, analyse, and evaluate vineyard output, enabling data-driven decisions to optimize yield and quality.

Key Sections

1. Yield per Block or Varietal

- **Block or Varietal Name**: [Insert name].
- **Harvest Date**: [Insert date].
- **Yield (tons/hectare or lbs/acre)**: [Insert value].
- **Total Weight of Grapes Collected**: [Insert weight].

2. Grape Quality Indicators

- **Sugar Levels (°Brix)**: [Insert value].
- **Acidity (Titratable Acidity in g/L)**: [Insert value].
- **Phenolic Content**:
    - Anthocyanin Levels: [Insert value].
    - Tannin Levels: [Insert value].
- **Other Indicators**:
    - pH: [Insert value].
    - Berry Size and Weight: [Insert observations].

3. Historical Yield Comparison

- **Current Season Yield**: [Insert yield].

- **Previous Season Yield**: [Insert yield].

- **Variance (% Increase/Decrease)**: [Insert percentage].

- **Key Factors Affecting Yield**:

    o Weather Conditions: [Insert notes].

    o Pest/Disease Incidence: [Insert notes].

    o Fertilization and Soil Management: [Insert observations].

4. Recommendations for Yield Optimization

- **Observations from Current Season**:

    o Successes: [Insert notes].

    o Challenges: [Insert notes].

- **Proposed Improvements**:

    o Irrigation Adjustments: [Insert details].

    o Fertilization Strategies: [Insert details].

    o Pest/Disease Management: [Insert details].

    o Canopy Management Techniques: [Insert details].

- **Action Plan for Next Season**:

    o Target Yield per Block/Varietal: [Insert target].

    o Planned Interventions: [Insert steps].

Additional Notes and Observations

[Insert any additional insights, comments, or observations related to vineyard productivity and yield.]

Marketing and Sales Plan Template

Purpose

To develop a strategic approach for effectively marketing and selling grapes or wine, ensuring profitability and customer engagement.

Key Sections

1. Target Markets and Customer Segments

- **Primary Target Markets**:
    - Local: [Insert details, e.g., nearby towns, regions].
    - National: [Insert details, e.g., urban areas, specialty markets].
    - International (if applicable): [Insert details, e.g., export markets].

- **Customer Segments**:
    - Wine Enthusiasts: [Insert profile].
    - Restaurants and Retailers: [Insert profile].
    - Farm-to-Table Consumers: [Insert profile].
    - Wholesale Buyers: [Insert profile].

- **Demographic Insights**:
    - Age Range: [Insert details].
    - Income Level: [Insert details].
    - Lifestyle Preferences: [Insert details].

2. Distribution Channels

- **Direct-to-Consumer (DTC)**:

- o Tasting Room Sales: [Insert strategy and target goals].
- o Online Store/E-Commerce: [Insert platform and target audience].
- **Wholesale and Retail**:
  - o Local Retail Partners: [Insert list and terms].
  - o Regional Distributors: [Insert details and agreements].
- **Export Opportunities**:
  - o Target Countries: [Insert regions].
  - o Compliance with Export Regulations: [Insert notes].

3. Pricing Strategies

- **Price Points**:
  - o Entry-Level Products: [Insert price range].
  - o Mid-Tier Products: [Insert price range].
  - o Premium or Reserve Lines: [Insert price range].
- **Cost Analysis**:
  - o Production Costs: [Insert details].
  - o Profit Margins: [Insert target percentages].
- **Discounts and Incentives**:
  - o Bulk Purchase Discounts: [Insert details].
  - o Seasonal Promotions: [Insert details].

4. Promotional Activities

- **Events and Tastings**:
  - o Wine Festivals: [Insert event names and dates].

- o   In-House Tastings: [Insert details].

- o   Collaboration with Local Restaurants: [Insert details].

- **Digital Marketing**:

  - o   Social Media Campaigns: [Insert platforms and goals].

  - o   Email Newsletters: [Insert frequency and content strategy].

  - o   Search Engine Optimization (SEO): [Insert focus keywords].

- **Farm-to-Table Partnerships**:

  - o   Collaborations with Chefs: [Insert names and focus].

  - o   Local Market Presence: [Insert target venues].

- **Traditional Marketing**:

  - o   Print Ads: [Insert publications].

  - o   Billboards or Signage: [Insert locations].

  - o   Local Radio/TV Spots: [Insert schedule].

5. Goals and Metrics

- **Sales Targets**:

  - o   Monthly Revenue: [Insert goals].

  - o   Units Sold: [Insert numbers].

- **Customer Acquisition**:

  - o   New Customers: [Insert goals].

  - o   Repeat Customers: [Insert goals].

- **Marketing ROI**:

  - o   Event Attendance and Conversions: [Insert metrics].

o   Online Engagement: [Insert goals, e.g., website traffic or social media interactions].

6. Budget and Resource Allocation

- **Marketing Budget**: [Insert total].

- **Allocation by Channel**:

  o   Digital Marketing: [Insert percentage].

  o   Events and Tastings: [Insert percentage].

  o   Traditional Media: [Insert percentage].

- **Staff and Vendor Costs**:

  o   Marketing Team: [Insert details].

  o   External Agencies or Consultants: [Insert details].

7. Monitoring and Adjustments

- **Regular Reviews**: Schedule [Insert frequency, e.g., monthly or quarterly] to evaluate sales and marketing performance.

- **Adapt Strategies**: Based on performance metrics and market feedback, adjust the approach as needed.

Risk Management and Contingency Plan Template

Purpose

To identify potential risks in vineyard operations, implement strategies to mitigate them, and establish response protocols to ensure business continuity during unexpected events.

Key Sections

1. Risk Identification

- **Weather-Related Risks**:

  o Frost, drought, excessive rainfall, hailstorms, heatwaves, or flooding.

  o Potential impact on grape health, yield, and vineyard infrastructure.

- **Pests and Diseases**:

  o Common vineyard pests: [Insert pests].

  o Common vineyard diseases: [Insert diseases, e.g., powdery mildew, botrytis].

- **Financial Risks**:

  o Cash flow shortages, market fluctuations, and unexpected operational costs.

  o Dependency on single buyers or markets.

- **Operational Risks**:

  o Equipment failures or shortages in labour.

  o Compliance issues with regulatory standards.

- **Environmental and Social Risks**:

o Impact of climate change and pressure from community or environmental groups.

2. Mitigation Strategies and Preventive Measures

- **Weather Preparedness**:
    - o Install frost protection systems (wind machines, heaters).
    - o Use rain covers or shade nets during extreme conditions.
    - o Monitor weather forecasts and adjust vineyard practices accordingly.

- **Pest and Disease Control**:
    - o Integrated Pest Management (IPM) strategies to reduce chemical dependence.
    - o Regular monitoring and early intervention for outbreaks.
    - o Biodiversity strategies, such as cover crops, to support natural pest control.

- **Financial Safeguards**:
    - o Diversify revenue streams, such as wine sales, agritourism, or farm-to-table partnerships.
    - o Maintain a reserve fund for emergencies.
    - o Purchase crop insurance to cover unforeseen losses.

- **Operational Stability**:
    - o Maintain and inspect equipment regularly to reduce failure risks.
    - o Develop a reliable labour pool through training and seasonal hiring plans.

- **Sustainability Measures**:
    - o Implement eco-friendly practices to mitigate climate impact and align with consumer expectations.

o Engage with the community to address environmental concerns proactively.

3. Emergency Response Protocols

- **Weather-Related Events**:

    o Identify safe areas for workers during storms or extreme weather.

    o Develop a system to alert staff of immediate risks.

- **Pest or Disease Outbreaks**:

    o Immediate isolation of affected vineyard blocks.

    o Implement rapid treatment plans, including biological or chemical solutions as needed.

- **Financial Crises**:

    o Prioritize essential expenses and defer non-critical spending.

    o Communicate with stakeholders, including lenders and suppliers, to negotiate terms.

- **Operational Failures**:

    o Create a backup plan for critical machinery, such as securing rental equipment.

    o Cross-train staff to handle multiple tasks during labour shortages.

4. Recovery Plans

- **Crop Loss Recovery**:

    o Replant affected blocks with resilient grape varieties or rootstocks.

    o Apply for government assistance or grants for agricultural recovery.

- **Financial Resilience**:

- o Develop a phased repayment plan for debts incurred during crises.

- o Launch promotional campaigns to boost short-term revenue.

- **Community Engagement**:

  - o Share recovery plans with stakeholders and customers to build trust and support.

  - o Collaborate with neighbouring vineyards to share resources and knowledge.

- **Long-Term Adjustments**:

  - o Evaluate the root causes of risks and refine vineyard practices.

  - o Incorporate climate-resilient technologies and strategies.

Monitoring and Updates

- **Regular Risk Assessments**: Schedule [Insert frequency, e.g., quarterly] reviews to identify emerging risks.

- **Update Protocols**: Adjust mitigation strategies and emergency responses based on lessons learned.

- **Training and Awareness**: Conduct regular training sessions for staff to ensure familiarity with the risk management plan.

Vineyard Sustainability Report Template

Purpose

To communicate the vineyard's sustainability achievements, outline current practices, and establish future goals for enhancing environmental and social responsibility.

Key Sections

1. Executive Summary

Provide a brief overview of the vineyard's sustainability efforts, achievements, and goals. Highlight key initiatives and their impact on environmental, economic, and social aspects.

2. Summary of Sustainable Practices

Water Use and Management

- Describe irrigation methods, water conservation strategies, and recycling systems.

- Include data on water usage reductions achieved compared to previous years.

Energy Conservation

- Detail renewable energy installations, such as solar panels or wind turbines.

- Explain measures taken to reduce energy consumption, such as energy-efficient machinery or lighting.

Soil Health and Fertility

- Outline practices like composting, organic fertilizers, and reduced tillage.

- Include soil testing results and amendments applied to maintain soil quality.

Waste Reduction

- Summarize efforts to minimize waste, such as composting grape pomace or recycling materials.

- Describe packaging innovations, such as lightweight bottles or biodegradable materials.

3. Biodiversity Preservation Efforts

Habitat Conservation

- Discuss actions taken to protect and enhance local ecosystems, such as creating wildlife corridors or planting native species.

Integrated Pest Management (IPM)

- Highlight practices that reduce chemical usage and promote natural pest control.

- Include data on reduced pesticide applications and the adoption of biological controls.

Vineyard Buffer Zones

- Explain the establishment of buffer zones to protect nearby waterways or sensitive habitats.

Pollinator Support

- Describe initiatives like planting wildflowers or maintaining beehives to support pollinators critical to vineyard health.

4. Carbon Footprint Analysis

Emissions Overview

- Present data on the vineyard's carbon emissions from operations, transportation, and energy use.

Reduction Initiatives

- Outline measures implemented to lower emissions, such as electrification of machinery or renewable energy use.

Offsets and Neutrality

- Detail carbon offset projects or partnerships contributing to the vineyard's carbon neutrality goals.

5. Goals for Improving Sustainability Performance

Short-Term Goals

- Specify actionable objectives for the upcoming year, such as further reductions in water or energy usage.

Long-Term Goals

- Define broader targets for the next 3–5 years, including achieving certifications, expanding renewable energy adoption, or reaching carbon neutrality.

Monitoring and Reporting

- Explain how progress will be tracked and reported, including planned updates to stakeholders.

6. Stakeholder Engagement and Education

- Discuss outreach efforts, including partnerships with local communities, consumer education, or collaborations with sustainability organizations.

7. Appendices and Supporting Data

- Include charts, graphs, and detailed metrics on sustainability achievements.
- Attach certifications or audit reports to validate the vineyard's sustainability claims.

Appendix C

Calendar of Key Vineyard Tasks

Annual Calendar of Key Vineyard Tasks for a Small-Scale Sustainable Vineyard

Winter Dormancy

- **Pruning**: Perform dormant pruning to shape vines and remove excess growth, ensuring balanced production for the coming season.

- **Vineyard Maintenance**: Repair trellis systems, stakes, and wires; clean and sanitize tools.

- **Soil Testing and Amendments**: Conduct soil tests and apply compost or organic amendments to replenish nutrients.

Late Winter

- **Disease Management**: Apply dormant sprays, such as lime sulphur, to control overwintering pests and diseases.

- **Planting Preparation**: Prepare new blocks or replace damaged vines; finalize orders for new vines.

- **Equipment Checks**: Inspect and service tractors, sprayers, and irrigation systems in preparation for spring.

Early Spring

- **Budbreak Monitoring**: Watch for signs of budbreak; protect vines from late frost using frost fans or other measures.

- **Irrigation Setup**: Test and adjust drip irrigation systems; monitor soil moisture levels.

- **Cover Crop Maintenance**: Mow or incorporate cover crops into the soil to improve organic matter and manage weeds.

Spring Growth

- **Shoot Thinning**: Remove excess shoots to promote airflow and optimize fruit exposure.

- **Weed Control**: Apply mulch or practice mechanical weed control to minimize competition.

- **Pest Monitoring**: Begin regular scouting for pests such as mites or aphids; implement Integrated Pest Management (IPM) strategies.

Late Spring

- **Canopy Management**: Perform early-season canopy work, such as positioning shoots or leaf removal, to optimize sunlight exposure.

- **Fertilization**: Apply organic fertilizers based on soil and tissue test results.

- **Flowering Management**: Monitor vine flowering and take steps to protect against adverse weather conditions.

Early Summer

- **Fruit Set Monitoring**: Assess fruit set and thin clusters as needed to balance vine load.

- **Disease Control**: Maintain a regular spray schedule with organic or biodynamic treatments to prevent mildew and other diseases.

- **Irrigation Management**: Adjust irrigation schedules to meet the increased water needs of developing vines.

Mid-Summer

- **Veraison Monitoring**: Observe the start of veraison (colour change in berries) and adjust canopy for even ripening.
- **Cluster Thinning**: Remove excess or uneven clusters to improve fruit quality.
- **Soil Health Practices**: Apply compost or organic mulch to retain moisture and suppress weeds.

Late Summer

- **Pre-Harvest Preparation**: Clean and sanitize harvest bins, tools, and equipment.
- **Grape Sampling**: Begin measuring sugar levels (Brix), acidity, and pH to determine optimal harvest timing.
- **Irrigation Reduction**: Gradually reduce water supply to concentrate flavours in grapes.

Early Fall/Autumn

- **Harvesting**: Hand-pick or use small-scale harvesters to collect grapes at peak ripeness.
- **Transport Logistics**: Ensure smooth transportation of grapes to the winery for immediate processing.
- **Post-Harvest Canopy Work**: Remove unripe clusters or damaged leaves to prevent disease spread.

Post-Harvest

- **Vineyard Cleanup**: Remove fallen leaves and debris to minimize disease pressure.
- **Cover Crop Seeding**: Plant winter cover crops to prevent erosion and improve soil structure.

- **Soil Sampling**: Collect post-harvest soil samples to plan amendments for winter.

Late Fall

- **Irrigation System Winterization**: Drain and protect irrigation lines from freezing.

- **Pruning Preparation**: Begin pre-pruning activities, such as removing dead wood.

- **Pest and Disease Assessment**: Scout for any lingering pest or disease issues and address them before dormancy.

Winter Rest

- **Tool Maintenance**: Sharpen and repair pruning shears and other equipment.

- **Planning and Record Keeping**: Review the year's vineyard performance; update records and refine next season's plan.

- **Rest and Recharge**: Allow the vineyard and team to rest in preparation for the next cycle.

References

1. Mendoza, G.H., et al., *Assessing Suitable Areas of Common Grapevine (Vitis Vinifera L.) for Current and Future Climate Situations: The CDS Toolbox SDM.* Atmosphere, 2020. **11**(11): p. 1201.

2. Fischer, N. and T. Efferth, *The Impact of "Omics" Technologies for Grapevine (Vitis Vinifera) Research.* Journal of Berry Research, 2021. **11**(4): p. 567-581.

3. Čeryová, N., et al., *Phenolic Contents, Antioxidant Activity and Colour Density of Slovak Pinot Noir Wines.* Acta Agriculturae Slovenica, 2021. **117**(3).

4. Gao, F., et al., *Preliminary Characterization of Chemical and Sensory Attributes for Grapes and Wines of Different Cultivars From the Weibei Plateau Region in China.* Food Chemistry X, 2024. **21**: p. 101091.

5. Terpou, A., et al., *Sustainable Solutions for Mitigating Spring Frost Effects on Grape and Wine Quality: Facilitating Digital Transactions in the Viniculture Sector.* Sustainable Food Technology, 2024. **2**(4): p. 967-975.

6. Mainar-Toledo, M.D., et al., *A Multi-Criteria Approach to Evaluate Sustainability: A Case Study of the Navarrese Wine Sector.* Energies, 2023. **16**(18): p. 6589.

7. Visconti, F., R. López, and M.Á. Olego, *The Health of Vineyard Soils: Towards a Sustainable Viticulture.* Horticulturae, 2024. **10**(2): p. 154.

8. Ederra, D.M., et al., *Connection Matters: Exploring the Implications of Scion–rootstock Alignment in Grafted Grapevines.* Australian Journal of Grape and Wine Research, 2022. **28**(4): p. 561-571.

9. Webb, L., P. Whetton, and E.W. Barlow, *Climate Change and Winegrape Quality in Australia.* Climate Research, 2008. **36**: p. 99-111.

10. White, R.E., *The Value of Soil Knowledge in Understanding Wine Terroir.* Frontiers in Environmental Science, 2020. **8**.

11. Cheng, G., et al., *Effects of Climatic Conditions and Soil Properties on Cabernet Sauvignon Berry Growth and Anthocyanin Profiles.* Molecules, 2014. **19**(9): p. 13683-13703.

12. Darriaut, R., et al., *Grapevine Rootstock and Soil Microbiome Interactions: Keys for a Resilient Viticulture.* Horticulture Research, 2022. **9**.

13. Bramley, R.G.V. and R. Hamilton, *Terroir and Precision Viticulture: Are They Compatible ?* Oeno One, 2007. **41**(1): p. 1.

14. Fraga, H., R. Costa, and J.A. Santos, *Multivariate Clustering of Viticultural Terroirs in the Douro Winemaking Region.* Ciência E Técnica Vitivinícola, 2017. **32**(2): p. 142-153.

15. Verarou, V., et al., *Effect of the Plot Variability on the Qualitative and Quantitative Characteristics of the Berry's Skins and Seeds of Grape Cultivar Agiorgitiko (Vitis Vinifera L.).* Viticulture Data Journal, 2020. **2**.

16. Bodin, F. and R. Morlat, *Characterization of Viticultural Terroirs Using a Simple Field Model Based on Soil Depth I. Validation of the Water Supply Regime, Phenology and Vine Vigour, in the Anjou Vineyard (France).* Plant and Soil, 2006. **281**(1-2): p. 37-54.

17. Cornelis, v.L., et al., *Soil Type and Soil Preparation Influence Vine Development and Grape Composition Through Its Impact on Vine Water and Nitrogen Status.* E3s Web of Conferences, 2018. **50**: p. 01015.

18. Rességuier, L.d., et al., *Variability of Climate, Water and Nitrogen Status and Its Influence on Vine Phenology and Grape Composition Inside a Small Winegrowing Estate.* E3s Web of Conferences, 2018. **50**: p. 01016.

19. Stefanis, C., et al., *Terroir in View of Bibliometrics.* Stats, 2023. **6**(4): p. 956-979.

20. Anesi, A., et al., *Towards a Scientific Interpretation of the Terroir Concept: Plasticity of the Grape Berry Metabolome.* BMC Plant Biology, 2015. **15**(1).

21. Leeuwen, C.v., J.P. Roby, and L.d. Rességuier, *Soil-Related Terroir Factors: A Review.* Oeno One, 2018. **52**(2): p. 173-188.

22. Vitulo, N., et al., *Bark and Grape Microbiome of Vitis Vinifera: Influence of Geographic Patterns and Agronomic Management on Bacterial Diversity.* Frontiers in Microbiology, 2019. **9**.

23. Gómez-Míguez, M. and F.J. Heredia, *Effect of the Maceration Technique on the Relationships Between Anthocyanin Composition and Objective Color of Syrah Wines.* Journal of Agricultural and Food Chemistry, 2004. **52**(16): p. 5117-5123.

24. Cerbu, I.M., et al., *The Effect of Different Techniques of Maceration-Fermentation on the Phenolic Composition of Red Wines.* Journal of Applied Life Sciences and Environment, 2021. **54**(1): p. 63-69.

25. Ortega-Regules, A., et al., *Changes in Skin Cell Wall Composition During the Maturation of Four Premium Wine Grape Varieties.* Journal of the Science of Food and Agriculture, 2007. **88**(3): p. 420-428.

26. Ghanem, E., et al., *Predicting the Composition of Red Wine Blends Using an Array of Multicomponent Peptide-Based Sensors.* Molecules, 2015. **20**(5): p. 9170-9182.

27. Lončarić, A., et al., *Changes in Volatile Compounds During Grape Brandy Production From 'Cabernet Sauvignon' and 'Syrah' Grape Varieties.* Processes, 2022. **10**(5): p. 988.

28. Pineau, B., et al., *Contribution of Grape Skin and Fermentation Microorganisms to the Development of Red- And Black-Berry Aroma in Merlot Wines.* Oeno One, 2011. **45**(1): p. 27.

29. Cortiella, M.G.i., et al., *Impact of Berry Size at Harvest on Red Wine Composition: A Winemaker's Approach.* Journal of the Science of Food and Agriculture, 2019. **100**(2): p. 836-845.

30. Mikulič-Petkovšek, M., et al., *Effects of Partial Dehydration Techniques on the Metabolite Compositionin 'Refošk' Grape Berries and Wine*.* Turkish Journal of Agriculture and Forestry, 2017. **41**: p. 10-22.

31. Vuorinen, H., K. Määttä, and R. Törrönen, *Content of the Flavonols Myricetin, Quercetin, and Kaempferol in Finnish Berry Wines.* Journal of Agricultural and Food Chemistry, 2000. **48**(7): p. 2675-2680.

32. Baiano, A. and C. Terracone, *Varietal Differences Among the Phenolic Profiles and Antioxidant Activities of Seven Table Grape Cultivars Grown in the South of Italy Based on Chemometrics.* Journal of Agricultural and Food Chemistry, 2011. **59**(18): p. 9815-9826.

33. Neves, A.C.L., et al., *Effect of Addition of Commercial Grape Seed Tannins on Phenolic Composition, Chromatic Characteristics, and Antioxidant Activity of Red Wine.* Journal of Agricultural and Food Chemistry, 2010. **58**(22): p. 11775-11782.

34. Mollica, A., et al., *Phenolic Analysis and in Vitro Biological Activity of Red Wine, Pomace and Grape Seeds Oil Derived From Vitis Vinifera L. Cv. Montepulciano D'Abruzzo.* Antioxidants, 2021. **10**(11): p. 1704.

35. González-Royo, E., et al., *The Effect of Supplementation With Three Commercial Inactive Dry Yeasts on the Colour, Phenolic Compounds, Polysaccharides and Astringency of a Model Wine Solution and Red Wine.* Journal of the Science of Food and Agriculture, 2016. **97**(1): p. 172-181.

36. Liu, X., et al., *Influence of Berry Heterogeneity on Phenolics and Antioxidant Activity of Grapes and Wines: A Primary Study of the New Winegrape Cultivar Meili (Vitis Vinifera L.).* Plos One, 2016. **11**(3): p. e0151276.

37. Monagas, M.a., et al., *Time Course of the Colour of Young Red Wines From <i>Vitis Vinifera</I> L. During Ageing in Bottle.* International Journal of Food Science & Technology, 2006. **41**(8): p. 892-899.

38. Hu, X.-Z., et al., *Geographical Origin Traceability of Cabernet Sauvignon Wines Based on Infrared Fingerprint Technology Combined With Chemometrics.* Scientific Reports, 2019. **9**(1).

39. Jeremic, J., et al., *The Oxygen Consumption Kinetics of Commercial Oenological Tannins in Model Wine Solution and Chianti Red Wine.* Molecules, 2020. **25**(5): p. 1215.

40. Tian, B., et al., *Proteomic Analysis of Sauvignon Blanc Grape Skin, Pulp and Seed and Relative Quantification of Pathogenesis-Related Proteins.* Plos One, 2015. **10**(6): p. e0130132.

41. Miller, K.S., A. Oberholster, and D.E. Block, *Creation and Validation of a Reactor Engineering Model for Multiphase Red Wine Fermentations.* Biotechnology and Bioengineering, 2019. **116**(4): p. 781-792.

42. Tsakiris, A., et al., *Red Wine Making by Immobilized Cells and Influence on Volatile Composition.* Journal of Agricultural and Food Chemistry, 2004. **52**(5): p. 1357-1363.

43. Sancho-Galán, P., V. Palacios, and A. Jiménez-Cantizano, *Effect of Grape Over-Ripening and Its Skin Presence on White Wine Alcoholic Fermentation in a Warm Climate Zone.* Foods, 2021. **10**(7): p. 1583.

44. Filipe-Ribeiro, L.s., et al., *Reducing the Negative Effect on White Wine Chromatic Characteristics Due to the Oxygen Exposure During Transportation by the Deoxygenation Process.* Foods, 2021. **10**(9): p. 2023.

45. Iobbi, A. and E. Tomasino, *Adapting Polarized Projective Mapping to Investigate Fruitiness Aroma Perception of White Wines From Oregon.* Beverages, 2021. **7**(3): p. 46.

46. Fracassetti, D., et al., *Chemical Characterization and Volatile Profile of Trebbiano Di Lugana Wine: A Case Study.* Foods, 2020. **9**(7): p. 956.

47. Tomasino, E., M. Song, and C. Fuentes, *Odor Perception Interactions Between Free Monoterpene Isomers and Wine Composition of Pinot Gris Wines.* Journal of Agricultural and Food Chemistry, 2020. **68**(10): p. 3220-3227.

48. Callaghan, C.M., R.E. Leggett, and R.M. Levin, *A Comparison of the Antioxidants and Carbohydrates in Common Wines and Grape Juices.* Free Radicals and Antioxidants, 2016. **7**(1): p. 86-89.

49. Lukić, K., et al., *Phenolic and Aroma Changes of Red and White Wines During Aging Induced by High Hydrostatic Pressure.* Foods, 2020. **9**(8): p. 1034.

50. Bordiga, M., et al., *Identification and Characterization of Complex Bioactive Oligosaccharides in White and Red Wine by a Combination of Mass Spectrometry and Gas Chromatography.* Journal of Agricultural and Food Chemistry, 2012. **60**(14): p. 3700-3707.

51. Zhang, Y., et al., *The Mechanism About the Resistant Dextrin Improving Sensorial Quality of Rice Wine and Red Wine.* Journal of Food Processing and Preservation, 2017. **41**(6): p. e13281.

52.   Sancho-Galán, P., V. Palacios, and A. Jiménez-Cantizano, *Volatile Composition and Sensory Characterization of Dry White Wines Made With Overripe Grapes by Means of Two Different Techniques.* Foods, 2022. **11**(4): p. 509.

53.   Melo, L., et al., *Predicting Wine Consumption Based on Previous 'Drinking History' and Associated Behaviours.* Journal of Food Research, 2012. **1**(1): p. 79.

54.   Binati, R.L., W.J.F. Lemos, and S. Torriani, *Contribution of Non-<i>Saccharomyces</I> Yeasts to Increase Glutathione Concentration in Wine.* Australian Journal of Grape and Wine Research, 2021. **27**(3): p. 290-294.

55.   Kontogeorgos, N. and I.G. Roussis, *Research Note: Total Free Sulphydryls of Several White and Red Wines.* South African Journal of Enology and Viticulture, 2016. **35**(1).

56.   Culbert, J.A., et al., *Classification of Sparkling Wine Style and Quality by MIR Spectroscopy.* Molecules, 2015. **20**(5): p. 8341-8356.

57.   Roussis, I.G., M. Patrianakou, and A. Drossiadis, *Protection of Aroma Volatiles in a Red Wine With Low Sulphur Dioxide by a Mixture of Glutathione, Caffeic Acid and Gallic Acid.* South African Journal of Enology and Viticulture, 2016. **34**(2).

58.   Gawel, R., et al., *Taste and Textural Characters of Mixtures of Caftaric Acid and Grape Reaction Product in Model Wine.* Australian Journal of Grape and Wine Research, 2013. **20**(1): p. 25-30.

59.   Sancho-Galán, P., et al., *Physicochemical and Nutritional Characterization of Winemaking Lees: A New Food Ingredient.* Agronomy, 2020. **10**(7): p. 996.

60.   Puértolas, E., et al., *Experimental Design Approach for the Evaluation of Anthocyanin Content of Rosé Wines Obtained by Pulsed Electric Fields. Influence of Temperature and Time of Maceration.* Food Chemistry, 2011. **126**(3): p. 1482-1487.

61.   Barrio-Galán, R.D., et al., *Volatile and Non-Volatile Characterization of White and Rosé Wines From Different Spanish Protected Designations of Origin.* Beverages, 2021. **7**(3): p. 49.

62.   Pozo-Bayón, M.Á., et al., *Impact of Using Trepat and Monastrell Red Grape Varieties on the Volatile and Nitrogen Composition During the Manufacture of Rosé Cava Sparkling Wines.* LWT, 2010. **43**(10): p. 1526-1532.

63.   Culbert, J.A., et al., *Influence of Production Method on the Sensory Profile and Consumer Acceptance of Australian Sparkling White Wine Styles.* Australian Journal of Grape and Wine Research, 2017. **23**(2): p. 170-178.

64.   Vecchio, R., et al., *The Role of Production Process and Information on Quality Expectations and Perceptions of Sparkling Wines*. Journal of the Science of Food and Agriculture, 2018. **99**(1): p. 124-135.

65.   Heit, C.E., et al., *Osmoadaptation of Wine Yeast (<i>Saccharomyces Cerevisiae</I>) During Icewine Fermentation Leads to High Levels of Acetic Acid*. Journal of Applied Microbiology, 2018. **124**(6): p. 1506-1520.

66.   Ma, Y., Y. Xu, and K. Tang, *Aroma of Icewine: A Review on How Environmental, Viticultural, and Oenological Factors Affect the Aroma of Icewine*. Journal of Agricultural and Food Chemistry, 2021. **69**(25): p. 6943-6957.

67.   Miller, G., et al., *Impact of ≪i>Botrytis Cinerea</I> On Γ-Nonalactone Concentration: Analysis of New Zealand White Wines Using SIDA-SPE-GC-MS*. Oeno One, 2024. **58**(4).

68.   Leça, J.M., et al., *Impact of Indigenous Non-Saccharomyces Yeasts Isolated From Madeira Island Vineyards on the Formation of Ethyl Carbamate in the Aging of Fortified Wines*. Processes, 2021. **9**(5): p. 799.

69.   Gaspar, J.M., et al., *Is Sotolon Relevant to the Aroma of Madeira Wine Blends?* Biomolecules, 2019. **9**(11): p. 720.

70.   Milheiro, J., et al., *Port Wine: Production and Ageing*. 2021.

71.   Ferreira, A.C.S., J.C. Barbe, and A. Bertrand, *3-Hydroxy-4,5-Dimethyl-2(5<i>H</I>)-Furanone: a Key Odorant of the Typical Aroma of Oxidative Aged Port Wine*. Journal of Agricultural and Food Chemistry, 2003. **51**(15): p. 4356-4363.

72.   Miranda, A., et al., *Impact of Non-Saccharomyces Yeast Fermentation in Madeira Wine Chemical Composition*. Processes, 2023. **11**(2): p. 482.

73.   Annunziata, E., et al., *The Role of Organizational Capabilities in Attaining Corporate Sustainability Practices and Economic Performance: Evidence From Italian Wine Industry*. Journal of Cleaner Production, 2018. **171**: p. 1300-1311.

74.   Dodds, R., et al., *What Drives Environmental Sustainability in the New Zealand Wine Industry?* International Journal of Wine Business Research, 2013. **25**(3): p. 164-184.

75.   Hauck, K., G. Szolnoki, and E. Pabst, *Motivation Factors for Organic Wines. An Analysis From the Perspective of German Producers and Retailers*. Wine Economics and Policy, 2021. **10**(2): p. 61-74.

76.   Moon, C., M.A. Bonn, and M. Cho, *A Comparison of the Importance of Wine Supplier Quality Attributes for on-Premise and Off-Premise Wine Retail Establishments*. International Journal of Contemporary Hospitality Management, 2024. **36**(8): p. 2795-2823.

77. Carrero, I., R.F. Redondo, and M.E. Fabra, *Who Is Behind the Sustainable Purchase? The Sustainable Consumer Profile in Grocery Shopping in Spain.* International Journal of Consumer Studies, 2016. **40**(6): p. 643-651.

78. Esposito, B., et al., *Exploring Corporate Social Responsibility in the Italian Wine Sector Through Websites.* The TQM Journal, 2021. **33**(7): p. 222-252.

79. Bouchagier, P., K. Tsimpoukas, and P. Kaldis, *Inferior Crop Performance in Organic Vine and Olive Sector Due to the Poor Implementation of Quality Processes. The Case of Kefallinia.* SDRP Journal of Plant Science, 2018. **2**(3): p. 1-10.

80. Vogel, S., et al., *The Ancient Rural Settlement Structure in the Hinterland of Pompeii Inferred From Spatial Analysis and Predictive Modeling of <i>Villae Rusticae</I>.* Geoarchaeology, 2016. **31**(2): p. 121-139.

81. Graça, A., et al., *Using Sustainable Development Actions to Promote the Relevance of Mountain Wines in Export Markets.* Open Agriculture, 2017. **2**(1): p. 571-579.

82. Bonn, M.A., J.J. Cronin, and M. Cho, *Do Environmental Sustainable Practices of Organic Wine Suppliers Affect Consumers' Behavioral Intentions? The Moderating Role of Trust.* Cornell Hospitality Quarterly, 2015. **57**(1): p. 21-37.

83. Lamoureux, C., N. Barbier, and T. Bouzdine-Chameeva, *Managing Wine Tourism and Biodiversity: The Art of Ambidexterity for Sustainability.* Sustainability, 2022. **14**(22): p. 15447.

84. Valero, A., J.A. Howarter, and J.W. Sutherland, *Sustainable Wine Scoring System (SWSS): A Life Cycle Assessment (LCA) Multivariable Approach.* Bio Web of Conferences, 2019. **12**: p. 03016.

85. Steur, H.D., et al., *Drivers, Adoption, and Evaluation of Sustainability Practices in Italian Wine SMEs.* Business Strategy and the Environment, 2019. **29**(2): p. 744-762.

86. D'Eusanio, M., B.M. Tragnone, and L. Petti, *From Social Accountability 8000 (SA8000) to Social Organisational Life Cycle Assessment (SO-LCA): An Evaluation of the Working Conditions of an Italian Wine-Producing Supply Chain.* Sustainability, 2022. **14**(14): p. 8833.

87. Pomarici, E., R. Vecchio, and A. Mariani, *Wineries' Perception of Sustainability Costs and Benefits: An Exploratory Study in California.* Sustainability, 2015. **7**(12): p. 16164-16174.

88. Visconti, K., *Red, White and Green: Environmental Communication on Wine Bottle Labels From New York's Hudson River Region.* Corporate Communications an International Journal, 2021. **26**(4): p. 728-757.

89. Schmit, T.M., B.J. Rickard, and J.T. Taber, *Consumer Valuation of Environmentally Friendly Production Practices in Wines, Considering*

Asymmetric Information and Sensory Effects. Journal of Agricultural Economics, 2012. **64**(2): p. 483-504.

90. Capitello, R. and L. Sirieix, *Consumers' Perceptions of Sustainable Wine: An Exploratory Study in France and Italy.* Economies, 2019. **7**(2): p. 33.

91. Sogari, G., et al., *Consumer Attitude Towards Sustainable-Labelled Wine: An Exploratory Approach.* International Journal of Wine Business Research, 2015. **27**(4): p. 312-328.

92. Csiba-Herczeg, Á., R. Koteczki, and B. Eisinger Balassa, *Sustainability Trends in the Wine Industry: Cognitive Biases and Methodological Insights From a PRISMA Review.* Ecocycles, 2023. **9**(3): p. 90-102.

93. Spielmann, N., *Larger and Better.* International Journal of Wine Business Research, 2017. **29**(2): p. 178-194.

94. Zambon, I., et al., *Rethinking Sustainability Within the Viticulture Realities Integrating Economy, Landscape and Energy.* Sustainability, 2018. **10**(2): p. 320.

95. Corbo, C., L. Lamastra, and E. Capri, *From Environmental to Sustainability Programs: A Review of Sustainability Initiatives in the Italian Wine Sector.* Sustainability, 2014. **6**(4): p. 2133-2159.

96. Di Chiara, V., et al., *Collaborative Approach for Achieving Ambitious Sustainability Goals: The Prosecco Sustainability Project.* Sustainability, 2024. **16**(2): p. 583.

97. Bonn, M.A., et al., *Green Purchasing by Wine Retailers: Roles of Individual Values, Competences and Organizational Culture.* Cornell Hospitality Quarterly, 2020. **62**(3): p. 324-336.

98. Litskas, V.D., et al., *Sustainable Viticulture: First Determination of the Environmental Footprint of Grapes.* Sustainability, 2020. **12**(21): p. 8812.

99. Arias-Navarro, I., et al., *Environmental Sustainability in Vineyards Under a Protected Designation of Origin in View of the Implementation of Photovoltaic Solar Energy Plants.* Land, 2023. **12**(10): p. 1871.

100. Gonçalves, F., et al., *Soil Arthropods in the Douro Demarcated Region Vineyards: General Characteristics and Ecosystem Services Provided.* Sustainability, 2021. **13**(14): p. 7837.

101. Ferrara, C. and G.D. Feo, *Life Cycle Assessment Application to the Wine Sector: A Critical Review.* Sustainability, 2018. **10**(2): p. 395.

102. Epuran, G., et al., *Food Safety and Sustainability ? An Exploratory Approach at the Level of the Romanian Wine Production Companies.* WWW Amfiteatrueconomic Ro, 2018. **20**(47): p. 151.

103. Navarro, A., et al., *Eco-Innovation and Benchmarking of Carbon Footprint Data for Vineyards and Wineries in Spain and France.* Journal of Cleaner Production, 2017. **142**: p. 1661-1671.

104. Turek Rahoveanu, P. and M. Turek Rahoveanu, *Assessment of Climate Changes on the Wine Sector.* Annals of the University of Craiova - Agriculture Montanology Cadastre Series, 2023. **53**(1): p. 319-323.

105. Reinhardt, T. and Y. Ambrogio, *Geographical Indications and Sustainable Viticulture: Empirical and Theoretical Perspectives.* Sustainability, 2023. **15**(23): p. 16318.

106. Droulia, F. and I. Charalampopoulos, *Future Climate Change Impacts on European Viticulture: A Review on Recent Scientific Advances.* Atmosphere, 2021. **12**(4): p. 495.

107. Rességuier, L.d., et al., *Temperature Variability at Local Scale in the Bordeaux Area. Relations With Environmental Factors and Impact on Vine Phenology.* Frontiers in Plant Science, 2020. **11**.

108. Gambetta, G.A. and S.K. Kurtural, *Global Warming and Wine Quality: Are We Close to the Tipping Point?* Oeno One, 2021. **55**(3): p. 353-361.

109. Haddad, E.A., et al., *A Bad Year? Climate Variability and the Wine Industry in Chile.* Wine Economics and Policy, 2020. **9**(2): p. 23-35.

110. Teixeira, A.H.d.C., et al., *Characterization of the Wine Grape Thermohydrological Conditions in the Tropical Brazilian Growing Region: Long-Term and Future Assessments.* Isrn Agronomy, 2014. **2014**: p. 1-14.

111. Nesbitt, A., et al., *Impact of Recent Climate Change and Weather Variability on the Viability of UK Viticulture - Combining Weather and Climate Records With Producers' Perspectives.* Australian Journal of Grape and Wine Research, 2016. **22**(2): p. 324-335.

112. Lee, H., et al., *Climate Change, Wine, and Conservation.* Proceedings of the National Academy of Sciences, 2013. **110**(17): p. 6907-6912.

113. Lereboullet, A.-L., et al., *The Viticultural System and Climate Change: Coping With Long-Term Trends in Temperature and Rainfall in Roussillon, France.* Regional Environmental Change, 2013. **14**(5): p. 1951-1966.

114. Santos, J.A., et al., *Statistical Modelling of Grapevine Yield in the Port Wine Region Under Present and Future Climate Conditions.* International Journal of Biometeorology, 2010. **55**(2): p. 119-131.

115. Malheiro, A.C., et al., *Climate Change Scenarios Applied to Viticultural Zoning in Europe.* Climate Research, 2010. **43**(3): p. 163-177.

116. Galbreath, J., *Climate Change Response: Evidence From the Margaret River Wine Region of Australia.* Business Strategy and the Environment, 2012. **23**(2): p. 89-104.

117. Sadras, V.O. and M. Moran, *Elevated Temperature Decouples Anthocyanins and Sugars in Berries of Shiraz and Cabernet Franc.* Australian Journal of Grape and Wine Research, 2012. **18**(2): p. 115-122.

118. Škrab, D., et al., *Cluster Thinning and Vineyard Site Modulate the Metabolomic Profile of Ribolla Gialla Base and Sparkling Wines.* Metabolites, 2021. **11**(5): p. 331.

119. Gouveia, C.M., et al., *Modelling Past and Future Wine Production in the Portuguese Douro Valley.* Climate Research, 2011. **48**(2): p. 349-362.

120. Dunn, M., et al., *The Future Potential for Wine Production in Scotland Under High-End Climate Change.* Regional Environmental Change, 2017. **19**(3): p. 723-732.

121. Upton, E., *Climate Change and Water Governance: Decision Making for Individual Vineyard Owners in Global Wine Regions.* Frontiers in Climate, 2021. **3**.

122. Schultze, S.R., P. Sabbatini, and L. Liu, *Effects of a Warming Trend on Cool Climate Viticulture in Michigan, USA.* Springerplus, 2016. **5**(1).

123. Bernáth, S., et al., *Influence of Climate Warming on Grapevine (Vitis Vinifera L.) Phenology in Conditions of Central Europe (Slovakia).* Plants, 2021. **10**(5): p. 1020.

124. Tóth, J. and Z. Végvári, *Future of Winegrape Growing Regions in Europe.* Australian Journal of Grape and Wine Research, 2015. **22**(1): p. 64-72.

125. Grazia, D., et al., *Grapes, Wines, and Changing Times: A Bibliometric Analysis of Climate Change Influence.* Australian Journal of Grape and Wine Research, 2023. **2023**: p. 1-19.

126. García-Orenes, F., et al., *Organic Fertilization in Traditional Mediterranean Grapevine Orchards Mediates Changes in Soil Microbial Community Structure and Enhances Soil Fertility.* Land Degradation and Development, 2016. **27**(6): p. 1622-1628.

127. Lima, F.V.d., et al., *Nitrogen and Organic Fertilization on Grapevine Productivity in the Brazilian Semiarid Region.* Revista Caatinga, 2019. **32**(1): p. 121-130.

128. Pérez-Álvarez, E.P., et al., *Effect of Two Doses of Urea Foliar Application on Leaves and Grape Nitrogen Composition During Two Vintages.* Journal of the Science of Food and Agriculture, 2016. **97**(8): p. 2524-2532.

129. Charrier, G., et al., *Drought Will Not Leave Your Glass Empty: Low Risk of Hydraulic Failure Revealed by Long-Term Drought Observations in World's Top Wine Regions.* Science Advances, 2018. **4**(1).

130. Bavougian, C.M. and P.E. Read, *Mulch and Groundcover Effects on Soil Temperature and Moisture, Surface Reflectance, Grapevine Water Potential, and Vineyard Weed Management.* Peerj, 2018. **6**: p. e5082.

131. Melo, G.W.d., et al., *Black Oats (<i>Avena Strigosa</I> Schreb) Solubilize Rock Phosphate and Provide Phosphorus to the Successive Crop, Grapevine (<i>Vitis Labrusca</I> L. 'Red Niagara').* Acta Horticulturae, 2018(1217): p. 89-98.

132. Brataševec, K., et al., *Hydroxycinnamic Acids as Affected by Different Fertilization of Rebula Grapevines.* Journal of Plant Nutrition and Soil Science, 2015. **178**(6): p. 868-877.

133. Brunetto, G., et al., *Use of the SPAD-502 in Estimating Nitrogen Content in Leaves and Grape Yield in Grapevines in Soils With Different Texture.* American Journal of Plant Sciences, 2012. **03**(11): p. 1546-1561.

134. Tullio, L., H. Morais, and R. Yagi, *Nutrition, Yield and Quality of 'Niagara Rosada' Vine Fruits Using Cattle Slurry and Plastic Cover.* Revista Brasileira De Fruticultura, 2018. **40**(6).

135. Domingues, R.R., et al., *Properties of Biochar Derived From Wood and High-Nutrient Biomasses With the Aim of Agronomic and Environmental Benefits.* Plos One, 2017. **12**(5): p. e0176884.

136. Hatch, T.A., C.C. Hickey, and T.K. Wolf, *Cover Crop, Rootstock, and Root Restriction Regulate Vegetative Growth of Cabernet Sauvignon in a Humid Environment.* American Journal of Enology and Viticulture, 2011. **62**(3): p. 298-311.

137. Baram, S., et al., *Estimating Nitrate Leaching to Groundwater From Orchards: Comparing Crop Nitrogen Excess, Deep Vadose Zone Data-Driven Estimates, and HYDRUS Modeling.* Vadose Zone Journal, 2016. **15**(11): p. 1-13.

138. Šimanský, V.r. and O. Ložek, *Fertilization of Vine by a 5-Aminolevulinic Acid-Based Fertilizer and Its Profitability.* Journal of Central European Agriculture, 2013. **14**(1): p. 270-283.

139. Kiss, T., et al., *Incidence of GLMD-Like Symptoms on Grapevines Naturally Infected by Grapevine Pinot Gris Virus, Boron Content and Gene Expression Analysis of Boron Metabolism Genes.* Agronomy, 2021. **11**(6): p. 1020.

140. Anđelini, D., et al., *Biochar From Grapevine-Pruning Residues Is Affected by Grapevine Rootstock and Pyrolysis Temperature.* Sustainability, 2023. **15**(6): p. 4851.

141. Delval, L., F. Jonard, and M. Javaux, *Simultaneous in Situ Monitoring of Belowground, Stem and Relative Stomatal Hydraulic Conductances of Grapevine Demonstrates a Soil-Texture Specific Transpiration Control.* 2024.

142. André, F., et al., *High-Resolution Imaging of a Vineyard in South of France Using Ground-Penetrating Radar, Electromagnetic Induction and Electrical Resistivity Tomography.* Journal of Applied Geophysics, 2012. **78**: p. 113-122.

143. Nevison, C.D., et al., *Denitrification, Leaching, and River Nitrogen Export in the Community Earth System Model.* Journal of Advances in Modeling Earth Systems, 2016. **8**(1): p. 272-291.

144. Bonada, M., et al., *Impact of Elevated Temperature and Water Deficit on the Chemical and Sensory Profiles of Barossa Shiraz Grapes and Wines.* Australian Journal of Grape and Wine Research, 2015. **21**(2): p. 240-253.

145. Savoi, S., et al., *Transcriptome and Metabolite Profiling Reveals That Prolonged Drought Modulates the Phenylpropanoid and Terpenoid Pathway in White Grapes (Vitis Vinifera L.).* BMC Plant Biology, 2016. **16**(1).

146. Bernardo, S., et al., *Grapevine Abiotic Stress Assessment and Search for Sustainable Adaptation Strategies in Mediterranean-Like Climates. A Review.* Agronomy for Sustainable Development, 2018. **38**(6).

147. Silvestroni, O., et al., *Effects of Limited Irrigation Water Volumes on Near-Isohydric 'Montepulciano' Vines Trained to Overhead Trellis System.* Acta Physiologiae Plantarum, 2020. **42**(9).

148. Medrano, H., et al., *Improving Water Use Efficiency of Vineyards in Semi-Arid Regions. A Review.* Agronomy for Sustainable Development, 2014. **35**(2): p. 499-517.

149. Vyshkvarkova, E., et al., *Assessment of the Current and Projected Conditions of Water Availability in the Sevastopol Region for Grape Growing.* Agronomy, 2021. **11**(8): p. 1665.

150. Pereyra, G. and M. Ferrer, *New Challenges for Uruguayan Viticulture.* Agrociencia Uruguay, 2024. **27**(NE1): p. e1195.

151. Fraga, H., et al., *Integrated Analysis of Climate, Soil, Topography and Vegetative Growth in Iberian Viticultural Regions.* Plos One, 2014. **9**(9): p. e108078.

152. Knipper, K., et al., *Evapotranspiration Estimates Derived Using Thermal-Based Satellite Remote Sensing and Data Fusion for Irrigation Management in California Vineyards.* Irrigation Science, 2018. **37**(3): p. 431-449.

153. Zarrouk, O., et al., *Grape Ripening Is Regulated by Deficit Irrigation/Elevated Temperatures According to Cluster Position in the Canopy.* Frontiers in Plant Science, 2016. **7**.

154. Leeuwen, C.v., et al., *An Update on the Impact of Climate Change in Viticulture and Potential Adaptations.* Agronomy, 2019. **9**(9): p. 514.

155. Mosedale, J.R., et al., *Climate Change Impacts and Adaptive Strategies: Lessons From the Grapevine.* Global Change Biology, 2016. **22**(11): p. 3814-3828.

156. Hofmann, M., R.L. Lux, and H.R. Schultz, *Constructing a Framework for Risk Analyses of Climate Change Effects on the Water Budget of Differently Sloped Vineyards With a Numeric Simulation Using the Monte Carlo Method Coupled to a Water Balance Model.* Frontiers in Plant Science, 2014. **5**.

157. Teslić, N., et al., *Climatic Shifts in High Quality Wine Production Areas, Emilia Romagna, Italy, 1961-2015.* Climate Research, 2017. **73**(3): p. 195-206.

158. Xanke, J. and T. Liesch, *Quantification and Possible Causes of Declining Groundwater Resources in the Euro-Mediterranean Region From 2003 to 2020.* Hydrogeology Journal, 2022. **30**(2): p. 379-400.

159. Finco, A., et al., *Combining Precision Viticulture Technologies and Economic Indices to Sustainable Water Use Management.* Water, 2022. **14**(9): p. 1493.

160. Martínez-Moreno, A., et al., *Is Deficit Irrigation With Saline Waters a Viable Alternative for Winegrowers in Semiarid Areas?* Oeno One, 2022. **56**(1): p. 101-116.

161. Alsafadi, K., et al., *Future Scenarios of Bioclimatic Viticulture Indices in the Eastern Mediterranean: Insights Into Sustainable Vineyard Management in a Changing Climate.* 2023.

162. Chen, R., et al., *Analyses of Vineyard Microclimate in the Eastern Foothills of the Helan Mountains in Ningxia Region, China.* Sustainability, 2023. **15**(17): p. 12740.

163. Souza, C.R.d., et al., *Row Orientation Effects on Syrah Grapevine Performance During Winter Growing Season.* Revista Ceres, 2019. **66**(3): p. 184-190.

164. Martins, G., et al., *Correlation Between Water Activity (Aw) and Microbial Epiphytic Communities Associated With Grapes Berries.* Oeno One, 2020. **54**(1): p. 49-61.

165. Kobayashi, H., S. Suzuki, and T. Takayanagi, *Correlations Between Climatic Conditions and Berry Composition of 'Koshu' (Vitis Vinifera) Grape in Japan.* Journal of the Japanese Society for Horticultural Science, 2011. **80**(3): p. 255-267.

166. Asproudi, A., et al., *Bunch Microclimate Affects Carotenoids Evolution in Cv. Nebbiolo (V. Vinifera L.).* Applied Sciences, 2020. **10**(11): p. 3846.

167. Vršič, S., et al., *The Impact of Climatic Warming on Earlier Wine-Grape Ripening in Northeastern Slovenia.* Horticulturae, 2024. **10**(6): p. 611.

168. Wang, Y., et al., *Effects of Basal Defoliation on Wine Aromas: A Meta-Analysis.* Molecules, 2018. **23**(4): p. 779.

169. Marais, J., J.J. Hunter, and P.D. Haasbroek, *Effect of Canopy Microclimate, Season and Region on Sauvignon Blanc Grape Composition and Wine Quality.* South African Journal of Enology and Viticulture, 2017. **20**(1).

170. Bokulich, N.A., et al., *Microbial Biogeography of Wine Grapes Is Conditioned by Cultivar, Vintage, and Climate.* Proceedings of the National Academy of Sciences, 2013. **111**(1).

171. VanderWeide, J., et al., *Enhancement of Fruit Technological Maturity and Alteration of the Flavonoid Metabolomic Profile in Merlot (<i>Vitis Vinifera</I> L.) by Early Mechanical Leaf Removal.* Journal of Agricultural and Food Chemistry, 2018. **66**(37): p. 9839-9849.

172. Scheper, J., et al., *Biodiversity and Pollination Benefits Trade Off Against Profit in an Intensive Farming System.* Proceedings of the National Academy of Sciences, 2023. **120**(28).

173. Muñoz-Sáez, A., J. Kitzes, and A.M. Merenlender, *Bird-friendly Wine Country Through Diversified Vineyards.* Conservation Biology, 2020. **35**(1): p. 274-284.

174. Castañeda, L.E. and O. Barbosa, *Metagenomic Analysis Exploring Taxonomic and Functional Diversity of Soil Microbial Communities in Chilean Vineyards and Surrounding Native Forests.* Peerj, 2017. **5**: p. e3098.

175. Ouvrard, S., S.M. Jasimuddin, and A. Spiga, *Does Sustainability Push to Reshape Business Models? Evidence From the European Wine Industry.* Sustainability, 2020. **12**(6): p. 2561.

176. Baird, T., C.M. Hall, and P. Castka, *New Zealand Winegrowers Attitudes and Behaviours Towards Wine Tourism and Sustainable Winegrowing.* Sustainability, 2018. **10**(3): p. 797.

177. Li, T., et al., *The Grapevine Transcription Factor VvTGA8 Enhances Resistance to White Rot via the Salicylic Acid Signaling Pathway in Tomato.* Agronomy, 2023. **13**(12): p. 3054.

178. Djennane, S., et al., *A Single Resistance Factor to Solve Vineyard Degeneration Due to Grapevine Fanleaf Virus.* Communications Biology, 2021. **4**(1).

179. Jiao, B., et al., *Engineering CRISPR Immune Systems Conferring GLRaV-3 Resistance in Grapevine.* Horticulture Research, 2022. **9**.

180. Calonnec, A., et al., *The Reliability of Leaf Bioassays for Predicting Disease Resistance on Fruit: A Case Study on Grapevine Resistance to Downy and Powdery Mildew.* Plant Pathology, 2012. **62**(3): p. 533-544.

181. Maraš, V., et al., *Population Genetic Analysis in Old Montenegrin Vineyards Reveals Ancient Ways Currently Active to Generate Diversity in Vitis Vinifera.* Scientific Reports, 2020. **10**(1).

182. Yin, W., et al., *Overexpression of <i>VqWRKY31</I> Enhances Powdery Mildew Resistance in Grapevine by Promoting Salicylic Acid Signaling and Specific Metabolite Synthesis.* Horticulture Research, 2022. **9**.

183. Gao, M., et al., *Characterization of Erysiphe Necator-Responsive Genes in Chinese Wild Vitis Quinquangularis.* International Journal of Molecular Sciences, 2012. **13**(9): p. 11497-11519.

184. Asghari, S., et al., *Induction of Systemic Resistance to <i>Agrobacterium Tumefaciens</I> by Endophytic Bacteria in Grapevine.* Plant Pathology, 2020. **69**(5): p. 827-837.

185. Jung, S.M., et al., *Profiling of Disease-Related Metabolites in Grapevine Internode Tissues Infected With Agrobacterium Vitis.* The Plant Pathology Journal, 2016. **32**(6): p. 489-499.

186. Rimbaud, L., et al., *Mosaics, Mixtures, Rotations or Pyramiding: What Is the Optimal Strategy to Deploy Major Gene Resistance?* Evolutionary Applications, 2018. **11**(10): p. 1791-1810.

187. Pap, D., et al., *Identification of Two Novel Powdery Mildew Resistance Loci, Ren6 and Ren7, From the Wild Chinese Grape Species Vitis Piasezkii.* BMC Plant Biology, 2016. **16**(1).

188. Legay, G., et al., *Identification of Genes Expressed During the Compatible Interaction of Grapevine With Plasmopara Viticola Through Suppression Subtractive Hybridization (SSH).* European Journal of Plant Pathology, 2010. **129**(2): p. 281-301.

189. Peressotti, E., et al., *Breakdown of Resistance to Grapevine Downy Mildew Upon Limited Deployment of a Resistant Variety.* BMC Plant Biology, 2010. **10**(1).

190. Liu, W., et al., *VqMAPK3/VqMAPK6, VqWRKY33, and <i>VqNSTS3</I> Constitute a Regulatory Node in Enhancing Resistance to Powdery Mildew in Grapevine.* Horticulture Research, 2023. **10**(7).

191. Shekhawat, U.K.S., T.R. Ganapathi, and A.B. Hadapad, *Transgenic Banana Plants Expressing Small Interfering RNAs Targeted Against Viral Replication Initiation Gene Display High-Level Resistance to Banana Bunchy Top Virus Infection.* Journal of General Virology, 2012. **93**(8): p. 1804-1813.

192. Zhu, Y., et al., *The Transcription Factors <i>VaERF16</I> and <i>VaMYB306</I> Interact to Enhance Resistance of Grapevine to <i>Botrytis Cinerea</I> Infection.* Molecular Plant Pathology, 2022. **23**(10): p. 1415-1432.

193. Li, T.G., et al., *Genome-Wide Identification and Functional Analyses of the CRK Gene Family in Cotton Reveals GbCRK18 Confers Verticillium Wilt Resistance in Gossypium Barbadense.* Frontiers in Plant Science, 2018. **9**.

194. Zhang, Y., et al., *Identification of the Defense-Related Gene VdWRKY53 From the Wild Grapevine Vitis Davidii Using RNA Sequencing and Ectopic Expression Analysis in Arabidopsis.* Hereditas, 2019. **156**(1).

195. Hemmer, C., et al., *Nanobody-mediated Resistance to Grapevine Fanleaf Virus in Plants.* Plant Biotechnology Journal, 2017. **16**(2): p. 660-671.

196. Wang, X.Y., et al., *Crude Garlic Extract Significantly Inhibits Replication of Grapevine Viruses.* Plant Pathology, 2019. **69**(1): p. 149-158.

197. Czigány, S., et al., *Application of a Topographic Pedosequence in the Villány Hills for Terroir Characterization*. Hungarian Geographical Bulletin, 2020. **69**(3): p. 245-261.
198. Bambach, N., et al., *Inter-Annual Variability of Land Surface Fluxes Across Vineyards: The Role of Climate, Phenology, and Irrigation Management*. Irrigation Science, 2022. **40**(4-5): p. 463-480.
199. Jackson, R.S., *Vineyard Practice*. 2014: p. 143-306.
200. Castellanos, B.G., B. García-García, and J.G. García, *Evaluation of the Sustainability of Vineyards in Semi-Arid Climates: The Case of Southeastern Spain*. Agronomy, 2022. **12**(12): p. 3213.
201. Aivazidou, E. and N. Tsolakis, *A Water Footprint Review of Italian Wine: Drivers, Barriers, and Practices for Sustainable Stewardship*. Water, 2020. **12**(2): p. 369.
202. García García, J., et al., *Sustainability of Vine Cultivation in Arid Areas of Southeastern Spain Through Strategies Combining Controlled Deficit Irrigation and Selection of Monastrell Clones*. Agronomy, 2023. **13**(8): p. 2046.
203. Ranca, A., et al., *Potential of Cover Crops to Control Arthropod Pests in Organic Viticulture*. Bulletin of University of Agricultural Sciences and Veterinary Medicine Cluj-Napoca Horticulture, 2022. **79**(1).
204. Abidin, M.A.Z., M.N. Mahyuddin, and M.A.A.M. Zainuri, *Solar Photovoltaic Architecture and Agronomic Management in Agrivoltaic System: A Review*. Sustainability, 2021. **13**(14): p. 7846.
205. Toga, M.T., *Solar Powered Irrigation for Sustainable Development and Its Risk in Ethiopia*. 2020.
206. Sunny, F.A., et al., *Determinants and Impact of Solar Irrigation Facility (SIF) Adoption: A Case Study in Northern Bangladesh*. Energies, 2022. **15**(7): p. 2460.
207. Carroquino, J., et al., *Combined Production of Electricity and Hydrogen From Solar Energy and Its Use in the Wine Sector*. Renewable Energy, 2018. **122**: p. 251-263.
208. Boquete, L., et al., *Portable System for Temperature Monitoring in All Phases of Wine Production*. Isa Transactions, 2010. **49**(3): p. 270-276.
209. Palazzo, L.R., et al., *Towards a More Sustainable Viticulture*. 2024. **1**.
210. Bugade, V.S. and P.K. Katti, *Assessment of Hybrid Energy Sources*. International Journal of Innovative Technology and Exploring Engineering, 2019. **9**(2): p. 4715-4720.
211. Nagarajan, K. and A.J.G. David, *Self-Excited Asynchronous Generator With PV Array in Detained Autonomous Generation Systems*. International Journal of Power Electronics and Drive Systems (Ijpeds), 2023. **14**(1): p. 358.

212. Kan, H.J. and L.N. Tan, *The Influence of Wind–Solar Hybrid Generation System on Transmission Service Rate.* Advanced Materials Research, 2014. **1070-1072**: p. 1472-1476.

213. Rosato, A., A. Perrotta, and L. Maffei, *Commercial Small-Scale Horizontal and Vertical Wind Turbines: A Comprehensive Review of Geometry, Materials, Costs and Performance.* Energies, 2024. **17**(13): p. 3125.

214. Asnaz, M.K., *A Case Study: Small Scale Wind Turbine System Selection and Economic Viability.* International Journal of Energy Applications and Technologies, 2018. **5**(4): p. 161-167.

215. Jou, H.L., et al., *Operation Strategy for Grid-Tied DC-coupling Power Converter Interface Integrating Wind/Solar/Battery.* Iop Conference Series Earth and Environmental Science, 2017. **93**: p. 012062.

216. Ngo, Q.-V., Y. Chai, and T.-T. Nguyen, *The Maximum Power Point Tracking Based-Control System for Small-Scale Wind Turbine Using Fuzzy Logic.* International Journal of Electrical and Computer Engineering (Ijece), 2020. **10**(4): p. 3927.

217. Stathopoulos, T., et al., *Urban Wind Energy: Some Views on Potential and Challenges.* Journal of Wind Engineering and Industrial Aerodynamics, 2018. **179**: p. 146-157.

218. Zhang, W., C.D. Markfort, and F. Porté-Agel, *Wind-Turbine Wakes in a Convective Boundary Layer: A Wind-Tunnel Study.* Boundary-Layer Meteorology, 2012. **146**(2): p. 161-179.

219. Toscano, G., et al., *Pelleting Vineyard Pruning at Low Cost With a Mobile Technology.* Energies, 2018. **11**(9): p. 2477.

220. Şenilă, L., et al., *Characterization of Biobriquettes Produced From Vineyard Wastes as a Solid Biofuel Resource.* Agriculture, 2022. **12**(3): p. 341.

221. Zanetti, M., et al., *Vineyard Pruning Residues Pellets for Use in Domestic Appliances: A Quality Assessment According to the EN ISO 17225.* Journal of Agricultural Engineering, 2017. **48**(2): p. 99.

222. Chico-Santamarta, L., et al., *Physical Quality Changes During the Storage of Canola (Brassica Napus L.) Straw Pellets.* Applied Energy, 2012. **95**: p. 220-226.

223. Weremczuk, A., *The Energy Potential of Agricultural Biomass in the European Union.* Zeszyty Naukowe SGGW W Warszawie - Problemy Rolnictwa Światowego, 2023. **23**(4): p. 44-60.

224. Ahorsu, R., F. Medina, and M. Constantí, *Significance and Challenges of Biomass as a Suitable Feedstock for Bioenergy and Biochemical Production: A Review.* Energies, 2018. **11**(12): p. 3366.

225. Pérez-Rodríguez, C.P., et al., *Harnessing Residual Biomass as a Renewable Energy Source in Colombia: A Potential Gasification Scenario.* Sustainability, 2022. **14**(19): p. 12537.

226. Yang, S., et al., *Research on the Development Status of Biomass Energy Serving the Construction of Ecological Civilization: A Case Study in Henan Province, China.* Bioresources, 2022. **18**(1): p. 465-483.

227. Radenahmad, N., et al., *Evaluation of the Bioenergy Potential of Temer Musa: An Invasive Tree From the African Desert.* International Journal of Chemical Engineering, 2021. **2021**: p. 1-10.

228. Chandraratne, M.R. and A.G. Daful, *Recent Advances in Thermochemical Conversion Of Biomass.* 2022.

229. Mekonnen Mossissa, H., et al., *Opportunities and Challenges of Harnessing Biomass Wastes for Decentralized Heat and Energy Generation and Climate Mitigation via Fluidized-Bed Gasification Pathway.* 2024.

230. Farghally, H.M., F.H. Fahmy, and M. Elsayed, *Geothermal Hot Water and Space Heating System in Egypt.* Renewable Energy and Power Quality Journal, 2010. **1**(08): p. 1578-1585.

231. Anifantis, A.S., S. Pascuzzi, and G. Scarascia-Mugnozza, *Geothermal Source Heat Pump Performance for a Greenhouse Heating System: An Experimental Study.* Journal of Agricultural Engineering, 2016. **47**(3): p. 164-170.

232. Santos, A.F., P.D. Gaspar, and H.J.L.d. Souza, *Innovative Hybrid Geothermal/Air Heat System for Biogas Production.* Journal of Engineering Research, 2023. **3**(27): p. 2-12.

233. Ignjatović, N.Ć., et al., *Sustainable Modularity Approach to Facilities Development Based on Geothermal Energy Potential.* Applied Sciences, 2021. **11**(6): p. 2691.

234. Cervera-Vázquez, J., C. Montagud, and J.M. Corberán, *In Situ Optimization Methodology for the Water Circulation Pumps Frequency of Ground Source Heat Pump Systems: Analysis for Multistage Heat Pump Units.* Energy and Buildings, 2015. **88**: p. 238-247.

235. Wu, Z., et al., *Thermo-Economic Analysis of Composite District Heating Substation With Absorption Heat Pump.* Applied Thermal Engineering, 2020. **166**: p. 114659.

236. Morrone, P. and A. Algieri, *Integrated Geothermal Energy Systems for Small-Scale Combined Heat and Power Production: Energy and Economic Investigation.* Applied Sciences, 2020. **10**(19): p. 6639.

237. Deng, J., et al., *What Is the Main Difference Between Medium-Depth Geothermal Heat Pump Systems and Conventional Shallow-Depth Geothermal Heat Pump Systems? Field Tests and Comparative Study.* Applied Sciences, 2019. **9**(23): p. 5120.

238. Ge, Y., Y. Sen, and Z. Yang, *Study on Zoning Effect of Shallow Geothermal Energy Suitability Based on Structural Matrix Method.* Strategic Planning for Energy and the Environment, 2024: p. 589-608.

239. Hou, J., X. Luo, and L. Zhang, *Establishment of Evaluation Model for Shallow Geothermal Energy Resource Development Potential Based on Characteristic of Geotemperature.* Earth Sciences Research Journal, 2020. **24**(3): p. 317-325.

240. García-Casarejos, N., P. Gargallo, and J. Carroquino, *Introduction of Renewable Energy in the Spanish Wine Sector.* Sustainability, 2018. **10**(9): p. 3157.

241. Carroquino, J., N. García-Casarejos, and P. Gargallo, *Introducing Renewable Energy in Vineyards and Agricultural Machinery: A Way to Reduce Emissions and Provide Sustainability.* Wine Studies, 2018. **2**.

242. Tobo, Y., V.R. Ancha, and G.S. Tibba, *CFD Simulation and Optimization of Very Low Head Axial Flow Turbine Runner.* International Journal of Renewable Energy Development, 2015. **4**(3): p. 181-188.

243. Kulišić, B., T. Radić, and M. Njavro, *Agro-Pruning for Energy as a Link Between Rural Development and Clean Energy Policies.* Sustainability, 2020. **12**(10): p. 4240.

244. Simões, C.L., et al., *Environmental Analysis of the Valorization of Woody Biomass Residues: A Comparative Study With Vine Pruning Leftovers in Portugal.* Sustainability, 2023. **15**(20): p. 14950.

245. Liu, W., et al., *Understanding the Water–Food–Energy Nexus for Supporting Sustainable Food Production and Conserving Hydropower Potential in China.* Frontiers in Environmental Science, 2019. **7**.

246. Grigg, N., et al., *The Water–food–energy Nexus in Pakistan: A Biophysical and Socio-Economic Challenge.* Proceedings of the International Association of Hydrological Sciences, 2018. **376**: p. 9-13.

247. Hagen, W., *Organic Grape Growing Techniques.* 2024, WineMaker.

248. Parlevliet, G. and S. McCoy, *Organic grapes and wine: a guide to production. Rural Industries Research and Development Corporation, Department of Agriculture, Western Australia, 2001.* Bulletin. **4516**.

249. Munitz, S., A. Schwartz, and Y. Netzer, *Effect of Timing of Irrigation Initiation on Vegetative Growth, Physiology and Yield Parameters in Cabernet Sauvignon Grapevines.* Australian Journal of Grape and Wine Research, 2020. **26**(3): p. 220-232.

250. Drori, E., et al., *The Effect of Irrigation-Initiation Timing on the Phenolic Composition and Overall Quality of Cabernet Sauvignon Wines Grown in a Semi-Arid Climate.* Foods, 2022. **11**(5): p. 770.

251. Torres, N., R. Yu, and S.K. Kurtural, *Arbuscular Mycorrhizal Fungi Inoculation and Applied Water Amounts Modulate the Response of Young*

Grapevines to Mild Water Stress in a Hyper-Arid Season. Frontiers in Plant Science, 2021. **11**.

252. Azorín, P.R. and J.G. García, *The Productive, Economic, and Social Efficiency of Vineyards Using Combined Drought-Tolerant Rootstocks and Efficient Low Water Volume Deficit Irrigation Techniques Under Mediterranean Semiarid Conditions.* Sustainability, 2020. **12**(5): p. 1930.

253. Ferrara, G., et al., *Soil Management Systems: Effects on Soil Properties and Weed Flora.* South African Journal of Enology and Viticulture, 2015. **36**(1).

254. Brunetto, G., et al., *Contribution of Mineral N to Young Grapevine in the Presence or Absence of Cover Crops.* Journal of Soil Science and Plant Nutrition, 2017. **17**(3): p. 570-580.

255. Brunetto, G., et al., *Contribution of Nitrogen From Agricultural Residues of Rye to 'Niagara Rosada' Grape Nutrition.* Scientia Horticulturae, 2014. **169**: p. 66-70.

256. Moukarzel, R., et al., *AMF Community Diversity Promotes Grapevine Growth Parameters Under High Black Foot Disease Pressure.* Journal of Fungi, 2022. **8**(3): p. 250.

257. Chancia, R., et al., *Assessing Grapevine Nutrient Status From Unmanned Aerial System (UAS) Hyperspectral Imagery.* Remote Sensing, 2021. **13**(21): p. 4489.

258. Holland, T., et al., *Does Inoculation With Arbuscular Mycorrhizal Fungi Reduce Trunk Disease in Grapevine Rootstocks?* Horticulturae, 2019. **5**(3): p. 61.

259. Zaller, J.G., et al., *Herbicides in Vineyards Reduce Grapevine Root Mycorrhization and Alter Soil Microorganisms and the Nutrient Composition in Grapevine Roots, Leaves, Xylem Sap and Grape Juice.* Environmental Science and Pollution Research, 2018. **25**(23): p. 23215-23226.

260. Ambrosini, V.G., et al., *Effect of Arbuscular Mycorrhizal Fungi on Young Vines in Copper-Contaminated Soil.* Brazilian Journal of Microbiology, 2015. **46**(4): p. 1045-1052.

261. Brunetto, G., et al., *Nutrient Release During the Decomposition of Mowed Perennial Ryegrass and White Clover and Its Contribution to Nitrogen Nutrition of Grapevine.* Nutrient Cycling in Agroecosystems, 2011. **90**(3): p. 299-308.

262. Zehetner, F., et al., *Soil Organic Carbon and Microbial Communities Respond to Vineyard Management.* Soil Use and Management, 2015. **31**(4): p. 528-533.

263. Stevanato, P., et al., *Soil Biological and Biochemical Traits Linked to Nutritional Status in Grapevine.* Journal of Soil Science and Plant Nutrition, 2014(ahead): p. 0-0.

264. Mirás-Avalos, J.M., et al., *Agronomic Practices for Reducing Soil Erosion in Hillside Vineyards Under Atlantic Climatic Conditions (Galicia, Spain).* Soil Systems, 2020. **4**(2): p. 19.

265. Okur, N., et al., *A Comparison of Soil Quality and Yield Parameters Under Organic and Conventional Vineyard Systems in Mediterranean Conditions (West Turkey).* Biological Agriculture & Horticulture, 2015. **32**(2): p. 73-84.

266. Schirmel, J., et al., *Positive Effects of Organic Viticulture on Carabid Beetles Depend on Landscape and Local Habitat Conditions.* Annals of Applied Biology, 2022. **181**(2): p. 192-200.

267. Hendgen, M., et al., *Effects of Different Management Regimes on Microbial Biodiversity in Vineyard Soils.* Scientific Reports, 2018. **8**(1).

268. Fuente, M.d.l., et al., *Large-Scale Implementation of Sustainable Production Practices in the Priorat-Montsant Region.* Bio Web of Conferences, 2019. **15**: p. 01014.

269. Frem, M., et al., *Sustainable Viticulture of Italian Grapevines: Environmental Evaluation and Societal Cost Estimation Using EU Farm Accountancy Data Network Data.* Horticulturae, 2023. **9**(11): p. 1239.

270. Telak, L.J., P. Pereira, and I. Bogunović, *Soil Degradation Mitigation in Continental Climate in Young Vineyards Planted in Stagnosols.* International Agrophysics, 2021. **35**(4): p. 307-317.

271. Graça, A., et al., *High-Resolution Geomatics Tools: Sustainably Managing Commercial Vineyards to Address the UN SDGs.* Bio Web of Conferences, 2023. **56**: p. 01010.

272. Westover, F., *Compost Use in Vineyards.* 2019, Extension.

273. Litskas, V.D., et al., *Use of Winery and Animal Waste as Fertilizers to Achieve Climate Neutrality in Non-Irrigated Viticulture.* Agronomy, 2022. **12**(10): p. 2375.

274. Lieskovský, J. and P. Kenderessy, *Modelling the Effect of Vegetation Cover and Different Tillage Practices on Soil Erosion in Vineyards: A Case Study in Vráble (Slovakia) Using Watem/Sedem.* Land Degradation and Development, 2012. **25**(3): p. 288-296.

275. Francioli, D., et al., *Land Use and Seasonal Effects on a Mediterranean Soil Bacterial Community.* Journal of Soil Science and Plant Nutrition, 2014(ahead): p. 0-0.

276. Muneret, L., et al., *Organic Farming Expansion Drives Natural Enemy Abundance but Not Diversity in Vineyard-dominated Landscapes.* Ecology and Evolution, 2019. **9**(23): p. 13532-13542.

277. Muneret, L., et al., *Deployment of Organic Farming at a Landscape Scale Maintains Low Pest Infestation and High Crop Productivity Levels in Vineyards.* Journal of Applied Ecology, 2017. **55**(3): p. 1516-1525.

278. Biddoccu, M., et al., *Assessment of Long-Term Soil Erosion in a Mountain Vineyard, Aosta Valley (NW Italy)*. Land Degradation and Development, 2017. **29**(3): p. 617-629.

279. Gabechaya, V., et al., *Exploring the Influence of Diverse Viticultural Systems on Soil Health Metrics in the Northern Black Sea Region*. Soil Systems, 2023. **7**(3): p. 73.

280. Lieskovský, J. and P. Kenderessy, *Degradation of Traditional Vineyards in Slovakia by Abandonment and Soil Erosion: A Case-study of Vráble*. Land Degradation and Development, 2022. **34**(1): p. 98-108.

281. Balestrini, R., et al., *Cohorts of Arbuscular Mycorrhizal Fungi (AMF) In<i>Vitis Vinifera</I>, a Typical Mediterranean Fruit Crop*. Environmental Microbiology Reports, 2010. **2**(4): p. 594-604.

282. Geel, M.V., et al., *High Soil Phosphorus Levels Overrule the Potential Benefits of Organic Farming on Arbuscular Mycorrhizal Diversity in Northern Vineyards*. Agriculture Ecosystems & Environment, 2017. **248**: p. 144-152.

283. Marques, M.J., et al., *Analysing Perceptions Attitudes and Responses of Winegrowers About Sustainable Land Management in Central Spain*. Land Degradation and Development, 2015. **26**(5): p. 458-467.

284. Coll, P., et al., *Organic Viticulture and Soil Quality: A Long-Term Study in Southern France*. Applied Soil Ecology, 2011.

285. Reiff, J.M., et al., *Organic Farming and Cover-Crop Management Reduce Pest Predation in Austrian Vineyards*. Insects, 2021. **12**(3): p. 220.

286. Taibi, H.H.Y., et al., *The Influence of No-Till Farming on Durum Wheat Mycorrhization in a Semi-Arid Region: A Long-Term Field Experiment*. Journal of Agricultural Science, 2020. **12**(4): p. 77.

287. Carlos, C., et al., *Parasitoids Of<i>Lobesia Botrana</I>(Lepidoptera: Tortricidae) in the Douro Demarcated Region Vineyards and the Prospects for Enhancing Conservation Biological Control*. Bulletin of Entomological Research, 2022. **112**(5): p. 697-706.

288. Rusch, A., L. Delbac, and D. Thiéry, *Grape Moth Density in Bordeaux Vineyards Depends on Local Habitat Management Despite Effects of Landscape Heterogeneity on Their Biological Control*. Journal of Applied Ecology, 2017. **54**(6): p. 1794-1803.

289. Wilson, H.F. and K.M. Daane, *Review of Ecologically-Based Pest Management in California Vineyards*. Insects, 2017. **8**(4): p. 108.

290. Burgio, G., et al., *Habitat Management of Organic Vineyard in Northern Italy: The Role of Cover Plants Management on Arthropod Functional Biodiversity*. Bulletin of Entomological Research, 2016. **106**(6): p. 759-768.

291. Möth, S., et al., *Unexpected Effects of Local Management and Landscape Composition on Predatory Mites and Their Food Resources in Vineyards.* Insects, 2021. **12**(2): p. 180.

292. Franin, K., B. Barić, and G. Kuštera, *The Role of Ecological Infrastructure on Beneficial Arthropods in Vineyards.* Spanish Journal of Agricultural Research, 2016. **14**(1): p. e0303.

293. Reiff, J.M., et al., *Arthropods on Grapes Benefit More From Fungicide Reduction Than From Organic Farming.* Pest Management Science, 2023. **79**(9): p. 3271-3279.

294. Zanettin, G., et al., *Influence of Vineyard Inter-Row Groundcover Vegetation Management on Arthropod Assemblages in the Vineyards of North-Eastern Italy.* Insects, 2021. **12**(4): p. 349.

295. Winter, S., et al., *Effects of Vegetation Management Intensity on Biodiversity and Ecosystem Services in Vineyards: A Meta-analysis.* Journal of Applied Ecology, 2018. **55**(5): p. 2484-2495.

296. Longa, C.M.O., et al., *Soil Microbiota Respond to Green Manure in Organic Vineyards.* Journal of Applied Microbiology, 2017. **123**(6): p. 1547-1560.

297. Garinie, T., et al., *Adverse Effects of the Bordeaux Mixture Copper-based Fungicide on the Non-target Vineyard Pest <scp><i>Lobesia Botrana</I></Scp>.* Pest Management Science, 2024. **80**(9): p. 4790-4799.

298. Leach, H., et al., *Evaluating Integrated Pest Management Tactics for Spotted Lanternfly Management in Vineyards.* Pest Management Science, 2023. **79**(10): p. 3486-3492.

299. Fernández-González, M., et al., *Prediction of Biological Sensors Appearance With ARIMA Models as a Tool for Integrated Pest Management Protocols.* Annals of Agricultural and Environmental Medicine, 2015. **23**(1): p. 129-137.

300. Paredes, D., et al., *Landscape Simplification Increases Vineyard Pest Outbreaks and Insecticide Use.* Ecology Letters, 2020. **24**(1): p. 73-83.

301. Jedlicka, J.A., R. Greenberg, and D.K. Letourneau, *Avian Conservation Practices Strengthen Ecosystem Services in California Vineyards.* Plos One, 2011. **6**(11): p. e27347.

302. Wilson, H.F., et al., *Landscape Diversity and Crop Vigor Outweigh Influence of Local Diversification on Biological Control of a Vineyard Pest.* Ecosphere, 2017. **8**(4).

303. Judt, C., et al., *Diverging Effects of Landscape Factors and Inter-Row Management on the Abundance of Beneficial and Herbivorous Arthropods in Andalusian Vineyards (Spain).* Insects, 2019. **10**(10): p. 320.

304. León, M., et al., *Evaluation of Sown Cover Crops and Spontaneous Weed Flora as a Potential Reservoir of Black-Foot Pathogens in Organic Viticulture.* Biology, 2021. **10**(6): p. 498.
305. Díaz, G.A., et al., *Prevalence and Pathogenicity of Fungi Associated With Grapevine Trunk Diseases in Chilean Vineyards.* Ciencia E Investigación Agraria, 2013. **40**(2): p. 327-339.
306. López-Granados, F., et al., *Monitoring Vineyard Canopy Management Operations Using UAV-Acquired Photogrammetric Point Clouds.* Remote Sensing, 2020. **12**(14): p. 2331.
307. Ouyang, J., et al., *UAV and Ground-Based Imagery Analysis Detects Canopy Structure Changes After Canopy Management Applications.* Oeno One, 2020. **54**(4): p. 1093-1103.
308. Gispert, C., J.D. Kaplan, and P.E. Rolshausen, *Long-Term Benefits of Protecting Table Grape Vineyards Against Trunk Diseases in the California Desert.* Agronomy, 2020. **10**(12): p. 1895.
309. Martínez-Diz, M.d.P., et al., *Grapevine Pruning Time Affects Natural Wound Colonization by Wood-Invading Fungi.* Fungal Ecology, 2020. **48**: p. 100994.
310. Aguilar, M.O., et al., *Influence of Vintage, Geographic Location and Cultivar on the Structure of Microbial Communities Associated With the Grapevine Rhizosphere in Vineyards of San Juan Province, Argentina.* Plos One, 2020. **15**(12): p. e0243848.
311. Berlanas, C., et al., *The Fungal and Bacterial Rhizosphere Microbiome Associated With Grapevine Rootstock Genotypes in Mature and Young Vineyards.* Frontiers in Microbiology, 2019. **10**.
312. Martiniuk, J.T., et al., *Grape-Associated Fungal Community Patterns Persist From Berry to Wine on a Fine Geographical Scale.* Fems Yeast Research, 2023. **23**.
313. Gupta, V.V.S.R., et al., *Vineyard Soil Microbiome Composition Related to Rotundone Concentration in Australian Cool Climate 'Peppery' Shiraz Grapes.* Frontiers in Microbiology, 2019. **10**.
314. López-García, Á., et al., *Space and Vine Cultivar Interact to Determine the Arbuscular Mycorrhizal Fungal Community Composition.* Journal of Fungi, 2020. **6**(4): p. 317.
315. Hofstetter, V., et al., *What if Esca Disease of Grapevine Were Not a Fungal Disease?* Fungal Diversity, 2012. **54**(1): p. 51-67.
316. Vanga, B.R., et al., *DNA Metabarcoding Reveals High Relative Abundance of Trunk Disease Fungi in Grapevines From Marlborough, New Zealand.* BMC Microbiology, 2022. **22**(1).

317.   Miura, T., et al., *Is Microbial Terroir Related to Geographic Distance Between Vineyards?* Environmental Microbiology Reports, 2017. **9**(6): p. 742-749.

318.   Fredrikson, L., P.A. Skinkis, and E. Peachey, *Cover Crop and Floor Management Affect Weed Coverage and Density in an Establishing Oregon Vineyard.* Horttechnology, 2011. **21**(2): p. 208-216.

319.   Trigo-Córdoba, E., et al., *Influence of Cover Crop Treatments on the Performance of a Vineyard in a Humid Region.* Spanish Journal of Agricultural Research, 2015. **13**(4): p. e0907.

320.   Abad, J., et al., *Under-Vine Cover Crops: Impact on Weed Development, Yield and Grape Composition.* Oeno One, 2020. **54**(4): p. 975-983.

321.   Fernando, M., L. Hale, and A. Shrestha, *Do Native and Introduced Cover Crops Differ in Their Ability to Suppress Weeds and Reduce Seedbanks? A Case Study in a Table Grape Vineyard.* 2023.

322.   Steiner, M., J.B. Grace, and S. Bacher, *Biodiversity Effects on Grape Quality Depend on Variety and Management Intensity.* Journal of Applied Ecology, 2021. **58**(7): p. 1442-1454.

323.   Jordan, L.M., T. Björkman, and J.E.V. Heuvel, *Annual Under-Vine Cover Crops Did Not Impact Vine Growth or Fruit Composition of Mature Cool-Climate 'Riesling' Grapevines.* Horttechnology, 2016. **26**(1): p. 36-45.

324.   Guinjuan, J.R., et al., *Dynamics of <i>Cynodon Dactylon</I> and Weed Community Composition in Different Cover Crops in a Vineyard.* Weed Research, 2023. **63**(4): p. 261-269.

325.   Fraga, H., et al., *An Overview of Climate Change Impacts on European Viticulture.* Food and Energy Security, 2012. **1**(2): p. 94-110.

326.   Mills-Novoa, M., P. Pszczólkowski, and F. Meza, *The Impact of Climate Change on the Viticultural Suitability of Maipo Valley, Chile.* The Professional Geographer, 2016. **68**(4): p. 561-573.

327.   Babin, N., et al., *Vineyard-Specific Climate Projections Help Growers Manage Risk and Plan Adaptation in the Paso Robles AVA.* California Agriculture, 2022. **75**(3): p. 142-150.

328.   Cabrera-Pérez, C., et al., *Cover Crops Terminated With Roller-crimper to Manage <i>Cynodon Dactylon</I> and Other Weeds in Vineyards.* Pest Management Science, 2024. **80**(4): p. 2162-2169.

329.   Wu, Z., et al., *Effects of Late Winter Pruning at Different Phenological Stages on Vine Yield Components and Berry Composition in La Rioja, North-Central Spain.* Oeno One, 2017. **51**(4): p. 363.

330.   Buesa, I., et al., *Effect of Delaying Winter Pruning of Bobal and Tempranillo Grapevines on Vine Performance, Grape and Wine Composition.* Australian Journal of Grape and Wine Research, 2020. **27**(1): p. 94-105.

331. Mutawila, C., F. Halleen, and L. Mostert, *Optimisation of Time of Application Of<i>Trichoderma</I>biocontrol Agents for Protection of Grapevine Pruning Wounds.* Australian Journal of Grape and Wine Research, 2016. **22**(2): p. 279-287.

332. Netzer, Y., et al., *Forever Young? Late Shoot Pruning Affects Phenological Development, Physiology, Yield and Wine Quality of Vitis Vinifera Cv. Malbec.* Agriculture, 2022. **12**(5): p. 605.

333. Falginella, L., et al., *Effect of Early Cane Pruning on Yield Components, Grape Composition, Carbohydrates Storage and Phenology in ≪i>Vitis Vinifera</I> L. Cv. Merlot.* Oeno One, 2022. **56**(3): p. 19-28.

334. Epee, P.T.M., et al., *Characterising Retained Dormant Shoot Attributes to Support Automated Cane Pruning On<i>Vitis Vinifera</I>L. Cv. Sauvignon Blanc.* Australian Journal of Grape and Wine Research, 2022. **28**(3): p. 508-520.

335. Monteiro, A.I., et al., *Assessment of Bud Fruitfulness of Three Grapevine Varieties Grown in Northwest Portugal.* Oeno One, 2022. **56**(3): p. 385-395.

336. Poni, S., et al., *Double Cropping in Vitis Vinifera L. Pinot Noir: Myth or Reality?* Agronomy, 2020. **10**(6): p. 799.

337. Gatti, M., et al., *Calibrated, Delayed-Cane Winter Pruning Controls Yield and Significantly Postpones Berry Ripening Parameters In<i>Vitis Vinifera</I>L. Cv. Pinot Noir.* Australian Journal of Grape and Wine Research, 2018. **24**(3): p. 305-316.

338. Bindon, K., P.R. Dry, and B.R. Loveys, *The Interactive Effect of Pruning Level and Irrigation Strategy on Water Use Efficiency of Vitis Vinifera L. Cv. Shiraz.* South African Journal of Enology and Viticulture, 2016. **29**(2).

339. Pascual, M., et al., *Canopy Management in Rainfed Vineyards (Cv. Tempranillo) for Optimising Water Use and Enhancing Wine Quality.* Journal of the Science of Food and Agriculture, 2015. **95**(15): p. 3067-3076.

340. He, L., et al., *Differential Influence of Timing and Duration of Bunch Bagging on Volatile Organic Compounds in Cabernet Sauvignon Berries (<scp><i>Vitis Vinifera</I></Scp>L.).* Australian Journal of Grape and Wine Research, 2021. **28**(1): p. 75-85.

341. Torres, N., et al., *Optimal Ranges and Thresholds of Grape Berry Solar Radiation for Flavonoid Biosynthesis in Warm Climates.* Frontiers in Plant Science, 2020. **11**.

342. Martínez-Lüscher, J., L. Brillante, and S.K. Kurtural, *Flavonol Profile Is a Reliable Indicator to Assess Canopy Architecture and the Exposure of Red Wine Grapes to Solar Radiation.* Frontiers in Plant Science, 2019. **10**.

343. O'Brien, P., C. Collins, and R.D. Bei, *Leaf Removal Applied to a Sprawling Canopy to Regulate Fruit Ripening in Cabernet Sauvignon.* Plants, 2021. **10**(5): p. 1017.

344. Gatti, M., et al., *Interactions of Summer Pruning Techniques and Vine Performance in the White<i>Vitis Vinifera</I>cv. Ortrugo.* Australian Journal of Grape and Wine Research, 2014. **21**(1): p. 80-89.

345. Garofalo, S.P., et al., *Agronomic Responses of Grapevines to an Irrigation Scheduling Approach Based on Continuous Monitoring of Soil Water Content.* Agronomy, 2023. **13**(11): p. 2821.

346. Greer, D.H. and M.N. Weedon, *The Impact of High Temperatures on Vitis Vinifera Cv. Semillon Grapevine Performance and Berry Ripening.* Frontiers in Plant Science, 2013. **4**.

347. Asproudi, A., et al., *Grape Aroma Precursors in Cv. Nebbiolo as Affected by Vine Microclimate.* Food Chemistry, 2016. **211**: p. 947-956.

348. Palliotti, A., et al., *Changes in Vineyard Establishment and Canopy Management Urged by Earlier Climate-Related Grape Ripening: A Review.* Scientia Horticulturae, 2014. **178**: p. 43-54.

349. Girardello, R.C., et al., *Impact of Grapevine Red Blotch Disease on Cabernet Sauvignon and Merlot Wine Composition and Sensory Attributes.* Molecules, 2020. **25**(14): p. 3299.

350. Rätsep, R., et al., *Recovery of Polyphenols From Vineyard Pruning Wastes—Shoots and Cane of Hybrid Grapevine (Vitis Sp.) Cultivars.* Antioxidants, 2021. **10**(7): p. 1059.

351. Wang, R., et al., *Effects of Regulated Deficit Irrigation on the Growth and Berry Composition of Cabernet Sauvignon in Ningxia.* International Journal of Agricultural and Biological Engineering, 2019. **12**(6): p. 102-109.

352. Suter, B., et al., *Modeling Stem Water Potential by Separating the Effects of Soil Water Availability and Climatic Conditions on Water Status in Grapevine (Vitis Vinifera L.).* Frontiers in Plant Science, 2019. **10**.

353. Martínez-Vidaurre, J.M., et al., *Differences in Soil Water Holding Capacity and Available Soil Water Along Growing Cycle Can Explain Differences in Vigour, Yield, and Quality of Must and Wine in the DOCa Rioja.* Horticulturae, 2024. **10**(4): p. 320.

354. Reshef, N., N. Agam, and A. Fait, *Grape Berry Acclimation to Excessive Solar Irradiance Leads to Repartitioning Between Major Flavonoid Groups.* Journal of Agricultural and Food Chemistry, 2018. **66**(14): p. 3624-3636.

355. Tramontini, S., et al., *Impact of Soil Texture and Water Availability on the Hydraulic Control of Plant and Grape-Berry Development.* Plant and Soil, 2012. **368**(1-2): p. 215-230.

356. Ramos, M.C., G.V. Jones, and J. Yuste, *Phenology and Grape Ripening Characteristics of Cv Tempranillo Within the Ribera Del Duero Designation of Origin (Spain): Influence of Soil and Plot Characteristics.* European Journal of Agronomy, 2015. **70**: p. 57-70.

357. Naraboli, V.C., et al., *Soil and Plant Nutrient Indices as Tools to Evaluate Nitrogen Management in Grape Production.* Journal of Natural Resource Conservation and Management, 2022. **3**(2): p. 118-124.

358. Teixeira, A.H.d.C., J. Tonietto, and J.F. Leivas, *Large-Scale Water Balance Indicators for Different Pruning Dates of Tropical Wine Grape.* Pesquisa Agropecuária Brasileira, 2016. **51**(7): p. 849-857.

359. Smith, H.M.S., et al., *Genetic Identification of SNP Markers Linked to a New Grape Phylloxera Resistant Locus in Vitis Cinerea for Marker-Assisted Selection.* BMC Plant Biology, 2018. **18**(1).

360. Zhang, K., et al., *Effects of the Fertilizer and Water Management on Amino Acids and Volatile Components in Cabernet Sauvignon Grapes and Wines.* International Journal of Agricultural and Biological Engineering, 2024. **17**(1): p. 69-79.

361. Resco, P., et al., *Exploring Adaptation Choices for Grapevine Regions in Spain.* Regional Environmental Change, 2015. **16**(4): p. 979-993.

362. Shmuleviz, R., et al., *Temperature Affects Organic Acid, Terpene and Stilbene Metabolisms in Wine Grapes During Postharvest Dehydration.* Frontiers in Plant Science, 2023. **14**.

363. He, F., et al., *Biosynthesis of Anthocyanins and Their Regulation in Colored Grapes.* Molecules, 2010. **15**(12): p. 9057-9091.

364. Reshef, N., et al., *Stable QTL for Malate Levels in Ripe Fruit and Their Transferability Across<i>Vitis</I>species.* Horticulture Research, 2022. **9**.

365. Zhu, Y., et al., *Exploring the Sensory Properties and Preferences of Fruit Wines Based on an Online Survey and Partial Projective Mapping.* Foods, 2023. **12**(9): p. 1844.

366. Marques, C.B., et al., *Environmental Sustainability and Biodynamic Cultivation of Vitis Viniferas Grapes in the Serra Gaúcha Region, Brazil.* International Journal of Advanced Engineering Research and Science, 2021. **8**(6): p. 349-359.

367. Sportelli, M., et al., *Autonomous Mowing and Complete Floor Cover for Weed Control in Vineyards.* Agronomy, 2021. **11**(3): p. 538.

368. Ramos, I.J., J. Ribeiro, and D. Figueiredo, *Effects of Vineyard Agricultural Practices on the Diversity of Macroinvertebrates.* Bio Web of Conferences, 2019. **12**: p. 01004.

369. Gutiérrez-Gamboa, G., et al., *Potential Opportunities of Thinned Clusters in Viticulture: A Mini Review.* Journal of the Science of Food and Agriculture, 2021. **101**(11): p. 4435-4443.

370. Rouxinol, M.I., et al., *Wine Grapes Ripening: A Review on Climate Effect and Analytical Approach to Increase Wine Quality.* Applied Biosciences, 2023. **2**(3): p. 347-372.

371. Nonni, F., et al., *Sentinel-2 Data Analysis and Comparison With UAV Multispectral Images for Precision Viticulture.* Gi_forum, 2018. **1**: p. 105-116.

372. Novara, A., et al., *Soil Carbon Budget Account for the Sustainability Improvement of a Mediterranean Vineyard Area.* Agronomy, 2020. **10**(3): p. 336.

373. Budziak-Wieczorek, I., et al., *Evaluation of the Quality of Selected White and Red Wines Produced From Moravia Region of Czech Republic Using Physicochemical Analysis, FTIR Infrared Spectroscopy and Chemometric Techniques.* Molecules, 2023. **28**(17): p. 6326.

374. Dorin, B., et al., *Utilization of Unmanned Aerial Vehicles for Zonal Winemaking in Cool-Climate Riesling Vineyards.* Oeno One, 2022. **56**(3): p. 327-341.

375. Ferreira, C.M., et al., *Assessment of the Impact of Distinct Vineyard Management Practices on Soil Physico-Chemical Properties.* Air Soil and Water Research, 2020. **13**.

376. Niwano, Y., M. Tada, and M. Tsukada, *Antimicrobial Intervention by Photoirradiation of Grape Pomace Extracts via Hydroxyl Radical Generation.* Frontiers in Physiology, 2017. **8**.

377. Belém, C.d.S., et al., *Digestibility, Fermentation and Microbiological Characteristics of Calotropis Procera Silage With Different Quantities of Grape Pomace.* Ciência E Agrotecnologia, 2016. **40**(6): p. 698-705.

378. Sheng, S., H. Yu, and Z. Zhang, *Strategies of Valorization of Sludge From Wastewater Treatment.* Journal of Chemical Technology & Biotechnology, 2018. **93**(4): p. 936-944.

379. Yengong, F.L., et al., *Variability of Physiochemical Properties of Livestock Manure With Added Wood Shavings During Windrow Composting.* African Journal of Environmental Science and Technology, 2021. **15**(2): p. 117-123.

380. Fernández, M.J., et al., *Effects of Clarification and Filtration Processes on the Removal of Fungicide Residues in Red Wines (Var. Monastrell).* Journal of Agricultural and Food Chemistry, 2005. **53**(15): p. 6156-6161.

381. Rayess, Y.E., et al., *Analysis of Membrane Fouling During Cross-Flow Microfiltration of Wine.* Innovative Food Science & Emerging Technologies, 2012. **16**: p. 398-408.

382. Vilanova, M., et al., *Use of a PGU1 Recombinant Saccharomyces Cerevisiae Strain in Oenological Fermentations.* Journal of Applied Microbiology, 2000. **89**(5): p. 876-883.

383. Arévalo-Villena, M., et al., *Pectinases Yeast Production Using Grape Skin as Carbon Source*. New Biotechnology, 2009. **25**: p. S70-S71.

384. Umiker, N.L., et al., *REMOVAL OF<i>BRETTANOMYCES BRUXELLENSIS</i>FROM RED WINE USING MEMBRANE FILTRATION*. Journal of Food Processing and Preservation, 2012. **37**(5): p. 799-805.

385. Millet, V. and A. Lonvaud-Funel, *The Viable but Non-Culturable State of Wine Micro-Organisms During Storage*. Letters in Applied Microbiology, 2000. **30**(2): p. 136-141.

386. Jiménez-Moreno, N. and C.A.n. Azpilicueta, *The Development of Esters in Filtered and Unfiltered Wines That Have Been Aged in Oak Barrels*. International Journal of Food Science & Technology, 2005. **41**(2): p. 155-161.

387. Shi, H., et al., *Towards Continuous Wine Fining: Materials Characterisation and Crossflow Performance Testing of Polymer–bentonite Mixed Matrix Membranes*. Asia-Pacific Journal of Chemical Engineering, 2017. **13**(1).

388. Rayess, Y.E., et al., *Wine Clarification With Rotating and Vibrating Filtration (RVF): Investigation of the Impact of Membrane Material, Wine Composition and Operating Conditions*. Journal of Membrane Science, 2016. **513**: p. 47-57.

389. Rutto, L.K., Z. Mersha, and M. Nita, *Evaluation of Cultivars and Spray Programs for Organic Grape Production in Virginia*. Horttechnology, 2021. **31**(2): p. 166-173.

390. Toaldo, I.M., et al., *Bioactive Potential of Vitis Labrusca L. Grape Juices From the Southern Region of Brazil: Phenolic and Elemental Composition and Effect on Lipid Peroxidation in Healthy Subjects*. Food Chemistry, 2015. **173**: p. 527-535.

391. Maioli, F., et al., *A Methodological Approach to Assess the Effect of Organic, Biodynamic, and Conventional Production Processes on the Intrinsic and Perceived Quality of a Typical Wine: The Case Study of Chianti DOCG*. Foods, 2021. **10**(8): p. 1894.

392. Kaltbach, S.B.d.A., et al., *Juices From 'Bordô' and 'BRS Cora' Grapes Grown in an Organic Production System in the Serra Do Sudeste Region*. Pesquisa Agropecuária Brasileira, 2022. **57**.

393. Hasanaliyeva, G., et al., *Effect of Organic and Conventional Production Methods on Fruit Yield and Nutritional Quality Parameters in Three Traditional Cretan Grape Varieties: Results From a Farm Survey*. Foods, 2021. **10**(2): p. 476.

394. Picchi, M., et al., *≪p>The Influence of Conventional and Biodynamic Winemaking Processes on the Quality of Sangiovese Wine</P&gt.* International Journal of Wine Research, 2020. **Volume 12**: p. 1-16.

395. Streletskaya, N.A., J. Liaukonytė, and H.M. Kaiser, *Absence Labels: How Does Information About Production Practices Impact Consumer Demand?* Plos One, 2019. **14**(6): p. e0217934.

396. Alonso González, P. and E. Parga-Dans, *The Natural Wine Phenomenon and the Promise of Sustainability: Institutionalization or Radicalization?* Culture Agriculture Food and Environment, 2023. **45**(2): p. 45-54.

397. Vecchio, R., et al., *Why Consumers Drink Natural Wine? Consumer Perception and Information About Natural Wine.* Agricultural and Food Economics, 2021. **9**(1).

398. Visconti, K., *Isabelle Legeron: <i>Natural Wine: An Introduction to Organic and Biodynamic Wines Made Naturally</I> CICO Books, New York, 2020, 224 Pp., ISBN 978-1782498995, $16.39.* Journal of Wine Economics, 2022. **17**(4): p. 352-354.

399. Sidari, R., et al., *Wine Yeasts Selection: Laboratory Characterization and Protocol Review.* Microorganisms, 2021. **9**(11): p. 2223.

400. Qiao, Y., et al., *Contribution of Grape Skins and Yeast Choice on the Aroma Profiles of Wines Produced From Pinot Noir and Synthetic Grape Musts.* Fermentation, 2021. **7**(3): p. 168.

401. Tao, Y., J. García, and D.W. Sun, *Advances in Wine Aging Technologies for Enhancing Wine Quality and Accelerating Wine Aging Process.* Critical Reviews in Food Science and Nutrition, 2013. **54**(6): p. 817-835.

402. Nevares, I. and M.d.Á. Sanza, *Characterization of the Oxygen Transmission Rate of New-Ancient Natural Materials for Wine Maturation Containers.* Foods, 2021. **10**(1): p. 140.

403. Sanza, M.d.Á., V.F. Laurie, and I. Nevares, *Wine Evolution and Spatial Distribution of Oxygen During Storage in High-Density Polyethylene Tanks.* Journal of the Science of Food and Agriculture, 2014. **95**(6): p. 1313-1320.

404. Issa-Issa, H., et al., *Effect of Aging Vessel (Clay-Tinaja Versus Oak Barrel) on the Volatile Composition, Descriptive Sensory Profile, and Consumer Acceptance of Red Wine.* Beverages, 2021. **7**(2): p. 35.

405. Maioli, F., et al., *Monitoring of Sangiovese Red Wine Chemical and Sensory Parameters Along One-Year Aging in Different Tank Materials and Glass Bottle.* Acs Food Science & Technology, 2022. **2**(2): p. 221-239.

406. Rubio-Bretón, P., T. Garde-Cerdán, and J. Martínez, *Use of Oak Fragments During the Aging of Red Wines. Effect on the Phenolic, Aromatic, and Sensory Composition of Wines as a Function of the Contact Time With the Wood.* Beverages, 2018. **4**(4): p. 102.

407. Tsukada, M., et al., *Microbicidal Action of Photoirradiated Aqueous Extracts From Wine Lees.* Journal of Food Science and Technology, 2016. **53**(7): p. 3020-3027.

408. Nocera, A., J.M. Ricardo-da-Silva, and S. Canas, *Antioxidant Activity and Phenolic Composition of Wine Spirit Resulting From an Alternative Ageing Technology Using Micro-Oxygenation: A Preliminary Study.* Oeno One, 2020. **54**(3): p. 485-496.

409. Pinter, E., et al., *Circularity Study on PET Bottle-to-Bottle Recycling.* Sustainability, 2021. **13**(13): p. 7370.

410. Guerrero, J.G., et al., *Sustainable Glass Recycling Culture-Based on Semi-Automatic Glass Bottle Cutter Prototype.* Sustainability, 2021. **13**(11): p. 6405.

411. Srivastava, R., et al., *Grid Interactive Solar Powered Automated Bottling Plant Using Microcontroller.* International Journal of Advanced Engineering and Management, 2017. **2**(1): p. 9.

412. Mura, R., et al., *Achieving the Circular Economy Through Environmental Policies: Packaging Strategies for More Sustainable Business Models in the Wine Industry.* Business Strategy and the Environment, 2023. **33**(2): p. 1497-1514.

413. Ruggeri, G., et al., *No More Glass Bottles? Canned Wine and Italian Consumers.* Foods, 2022. **11**(8): p. 1106.

414. Buchanan, C., et al., *Lightweighting Shipping Containers: Life Cycle Impacts on Multimodal Freight Transportation.* Transportation Research Part D Transport and Environment, 2018. **62**: p. 418-432.

415. Valenzuela, L., et al., *Consumer Willingness to Pay for Sustainable Wine— The Chilean Case.* Sustainability, 2022. **14**(17): p. 10910.

416. Wagner, M., et al., *Developing a Sustainability Vision for the Global Wine Industry.* Sustainability, 2023. **15**(13): p. 10487.

417. Migliore, G., et al., *Factors Affecting Consumer Preferences for "Natural Wine".* British Food Journal, 2020. **122**(8): p. 2463-2479.

418. Rui, M., et al., *Understanding Factors Associated With Interest in Sustainability-Certified Wine Among American and Italian Consumers.* Foods, 2024. **13**(10): p. 1468.

419. Sogari, G., C. Mora, and D. Menozzi, *Factors Driving Sustainable Choice: The Case of Wine.* British Food Journal, 2016. **118**(3): p. 632-646.

420. Ignjatijević, S., et al., *Agro-Environmental Practices and Business Performance in the Wine Sector.* Agriculture, 2022. **12**(2): p. 239.

421. Barber, N., D. Taylor, and D. Remar, *Desirability Bias and Perceived Effectiveness Influence on Willingness-to-Pay for Pro-Environmental Wine Products.* International Journal of Wine Business Research, 2016. **28**(3): p. 206-227.

422. Gómez-Borja, M.-Á., I. Carrasco, and J.S. Castillo-Valero, *User-Generated Content and Relevance of Sustainability Dimensions in the Wine Market.* Bio Web of Conferences, 2023. **68**: p. 03019.

423. Stanco, M. and M. Lerro, *Consumers' Preferences for and Perception of CSR Initiatives in the Wine Sector.* Sustainability, 2020. **12**(13): p. 5230.

424. Moscovici, D., et al., *Preferences for Eco Certified Wines in the United States.* International Journal of Wine Business Research, 2020. **33**(2): p. 153-175.

425. Henley, C.D., et al., *Label Design: Impact on Millennials' Perceptions of Wine.* International Journal of Wine Business Research, 2011. **23**(1): p. 7-20.

426. Sogari, G., et al., *Millennial Generation and Environmental Sustainability: The Role of Social Media in the Consumer Purchasing Behavior for Wine.* Sustainability, 2017. **9**(10): p. 1911.

427. Nassivera, F., et al., *Italian Millennials' Preferences for Wine: An Exploratory Study.* British Food Journal, 2020. **122**(8): p. 2403-2423.

428. Muñoz, R.M., M.V. Fernández, and M.Y.S. Martín, *Assessing Consumer Behavior in the Wine Industry and Its Consequences for Wineries: A Case Study of a Spanish Company.* Frontiers in Psychology, 2019. **10**.

429. Alebaki, M., et al., *Digital Winescape and Online Wine Tourism: Comparative Insights From Crete and Santorini.* Sustainability, 2022. **14**(14): p. 8396.

430. Kieling, A.P., R. Tezza, and G.L. Vargas, *Website Stage Model for Brazilian Wineries: An Analysis of Presence in Digital and Mobile Media.* International Journal of Wine Business Research, 2022. **35**(1): p. 45-65.

431. Costopoulou, C., M. Ntaliani, and F. Ntalianis, *An Analysis of Social Media Usage in Winery Businesses.* Advances in Science Technology and Engineering Systems Journal, 2019. **4**(4): p. 380-387.

432. Thach, L., T. Lease, and M. Barton, *Exploring the Impact of Social Media Practices on Wine Sales in US Wineries.* Journal of Direct Data and Digital Marketing Practice, 2016. **17**(4): p. 272-283.

433. Bitakou, E., et al., *Evaluating Social Media Marketing in the Greek Winery Industry.* Sustainability, 2023. **16**(1): p. 192.

434. Dreßler, M. and I. Paunović, *A Typology of Winery SME Brand Strategies With Implications for Sustainability Communication and Co-Creation.* Sustainability, 2021. **13**(2): p. 805.

435. Ingrassia, M., et al., *The Wine Influencers: Exploring a New Communication Model of Open Innovation for Wine Producers—A Netnographic, Factor and AGIL Analysis.* Journal of Open Innovation Technology Market and Complexity, 2020. **6**(4): p. 165.

436. Perretti, B., *Economic Sustainability of Quality Wine Districts in the South of Italy. The Case of Vulture.* International Journal of Globalisation and Small Business, 2020. **1**(1): p. 1.

437. Borsellino, V., et al., *The Sicilian Cooperative System of Wine Production.* International Journal of Wine Business Research, 2020. **32**(3): p. 391-421.

438. Perovic, N., et al., *Analysis of the Promotion of Small Wine Producers in Wine Regions of Montenegro and the Perspective of Wine Tourism in Cooperation With Local Tourism Organizations.* The Journal Agriculture and Forestry, 2024. **70**(1).

439. Ollat, N., J.-M. Touzard, and C.v. Leeuwen, *Climate Change Impacts and Adaptations: New Challenges for the Wine Industry.* Journal of Wine Economics, 2016. **11**(1): p. 139-149.

440. Bandinelli, R., et al., *Environmental Practices in the Wine Industry: An Overview of the Italian Market.* British Food Journal, 2020. **122**(5): p. 1625-1646.

441. Casanova-Gascón, J., et al., *Behavior of Vine Varieties Resistant to Fungal Diseases in the Somontano Region.* Agronomy, 2019. **9**(11): p. 738.

442. Lubell, M., V. Hillis, and M. Hoffman, *Innovation, Cooperation, and the Perceived Benefits and Costs of Sustainable Agriculture Practices.* Ecology and Society, 2011. **16**(4).

443. Ambaye, M., *Small French Wineries' Export Strategies to China.* Journal of Management Research, 2015. **7**(2): p. 109.

444. Kelley, K.M., et al., *Identifying Wine Consumers Interested in Environmentally Sustainable Production Practices.* International Journal of Wine Business Research, 2021. **34**(1): p. 86-111.

445. Temple, C., et al., *The Impact of Changes in Regulatory and Market Environment on Sustainability of Wine Producers: A Structural Equation Model.* Wine Economics and Policy, 2020. **9**(1): p. 51-61.

446. Gázquez–Abad, J.C., et al., *Drivers of Sustainability Strategies in Spain's Wine Tourism Industry.* Cornell Hospitality Quarterly, 2014. **56**(1): p. 106-117.

447. Pero, C., et al., *IoT-Driven Machine Learning for Precision Viticulture Optimization.* Ieee Journal of Selected Topics in Applied Earth Observations and Remote Sensing, 2024. **17**: p. 2437-2447.

448. Sassu, A., et al., *Advances in Unmanned Aerial System Remote Sensing for Precision Viticulture.* Sensors, 2021. **21**(3): p. 956.

449. Brunori, E., et al., *Field Survey and UAV Remote Sensing as Tools for Evaluating the Canopy Management Effects in Smallholder Grapevine Farm.* Bio Web of Conferences, 2022. **44**: p. 05001.

450. Bastard, A. and A. Chaillet, *Digitalization From Vine to Wine: Successes and Remaining Challenges - A Review.* Bio Web of Conferences, 2023. **68**: p. 01034.

451. Pivac, T., et al., *The Importance of Digital Marketing for Wineries and Development of Wine Tourism: Case Study of Serbia.* 2020: p. 241-251.

452. Duchêne, É., *How Can Grapevine Genetics Contribute to the Adaptation to Climate Change?* Oeno One, 2016. **50**(3).

453. Li, Y. and I. Bardají, *Adapting the Wine Industry in China to Climate Change: Challenges and Opportunities.* Oeno One, 2017. **51**(2): p. 71-89.

454. Morande, J.A., et al., *From Berries to Blocks: Carbon Stock Quantification of a California Vineyard.* Carbon Balance and Management, 2017. **12**(1).

455. Villat, J. and K.A. Nicholas, *Quantifying Soil Carbon Sequestration From Regenerative Agricultural Practices in Crops and Vineyards.* Frontiers in Sustainable Food Systems, 2024. **7**.

456. Abbate, S., P. Centobelli, and M. Di Gregorio, *Wine Waste Valorisation: Crushing the Research Domain.* Review of Managerial Science, 2024.

457. Rodrigues, R., L.M. Gando-Ferreira, and M.J. Quina, *Increasing Value of Winery Residues Through Integrated Biorefinery Processes: A Review.* Molecules, 2022. **27**(15): p. 4709.

458. Francioli, D., et al., *Roots of Resilience: Optimizing Microbe-rootstock Interactions to Enhance Vineyard Productivity.* Plants People Planet, 2024.

459. Özbakır, O., *Safety Measures and Risk Management in Agricultural Confined Spaces: A Study on Farm in Iğdır Province, Using Bow Tie and Matrix Methods.* Journal of Agriculture, 2024. **7**(1): p. 31-44.

460. Burlet-Vienney, D., et al., *Occupational Safety During Interventions in Confined Spaces.* Safety Science, 2015. **79**: p. 19-28.

461. Siani, A.M., et al., *Occupational Exposures to Solar Ultraviolet Radiation of Vineyard Workers in Tuscany (Italy).* Photochemistry and Photobiology, 2011. **87**(4): p. 925-934.

462. Benito, S., *The Management of Compounds That Influence Human Health in Modern Winemaking From an HACCP Point of View.* Fermentation, 2019. **5**(2): p. 33.

463. Wilson, M.P., H.N. Madison, and S.B. Healy, *Confined Space Emergency Response: Assessing Employer and Fire Department Practices.* Journal of Occupational and Environmental Hygiene, 2012. **9**(2): p. 120-128.

464. Wilson, P. and Q. Wang, *Development of a Protocol for Determining Confined Space Occupant Load.* Process Safety Progress, 2013. **33**(2): p. 143-147.

465. Moreno-García, J., et al., *Yeast Immobilization Systems for Alcoholic Wine Fermentations: Actual Trends and Future Perspectives.* Frontiers in Microbiology, 2018. **9**.

466.   Andersen, J.H., et al., *Systematic Literature Review on the Effects of Occupational Safety and Health (OSH) Interventions At the Workplace.* Scandinavian Journal of Work Environment & Health, 2018. **45**(2): p. 103-113.

Index